Innovation and Technological Change

An International Comparison

Edited by
Zoltan J. Acs and David B. Audretsch

Ann Arbor
The University of Michigan Press

Published in the United States of America by
The University of Michigan Press

Library of Congress Cataloging-in-Publication Data

Innovation and technological change: an international comparison /
  edited by Zoltán J. Àcs and David B. Audretsch.
     p.   cm.
  Includes bibliographical references and index.
  ISBN 0-472-10249-4, (cloth)
  1. Technological innovations—Economic aspects—Case studies.
I. Àcs, Zoltán J.   II. Audretsch, David B.
HC79.T4I546 1991
338'.064—dc20

Printed in Great Britain

1994 1993 1992 1991 4 3 2 1

Contents

List of contributors

Zoltan J. Acs University of Baltimore

David B. Audretsch Wissenschaftszentrum Berlin für Sozialforschung

Alok K. Chakrabarti New Jersey Institute of Technology

Wesley M. Cohen Carnegie-Mellon University

Felix R. FitzRoy St Andrews University

Michael R. Halperin University of Pennsylvania

Alfred Kleinknecht University of Amsterdam

Steven Klepper Carnegie-Mellon University

Kornelius Kraft University of Kassel

Albert N. Link University of North Carolina – Greensboro

Tom P. Poot University of Amsterdam

John Rees University of North Carolina – Greensboro

Jeroen O. N. Reijnen University of Amsterdam

F. M. Scherer Harvard University

J.-Matthias Graf von der Schulenburg University of Hannover

Joachim Schwalbach Free University Berlin

John T. Scott Dartmouth College

Paul Stoneman University of Warwick

Joachim Wagner University of Hannover

Klaus F. Zimmermann University of Munich

Acknowledgments

The papers included in this volume are an edited selection of reports presented at the conference, "Innovation and technological change: An international comparison", held at the Wissenschaftszentrum Berlin für Sozialforschung on August 10–11, 1989.

The conference as well as this volume would not have been possible without the assistance of several key people. In particular, we would like to thank the German Marshall Fund of the United States and the Wissenschaftszentrum Berlin für Sozialforschung for financing the conference. Special thanks go to Manfred Fleischer, who helped with the institutional arrangements, as well as Linda Kronberg and Christiane Loycke de Roux, who were responsible for the conference organization and assembling the manuscripts. We are also grateful to Peter Johns of Harvester Wheatsheaf, who guided us from the conference stage to publication with as few impediments as possible.

1

Innovation and technological change: An overview

Zoltan J. Acs and David B. Audretsch

1.1   Introduction

That the subject of innovation and technological change has, within the span of just a few short years, been transformed from a somewhat arcane academic discipline to one of the most pressing issues of national debate is apparent from the current wave of attention in the popular press on both sides of the Atlantic. The reason underlying the emergence of innovation as a topic of broad public interest was explained by F. M. Scherer (1988, p. 1) to the US Congress: "The question of innovation is particularly important, for upon the innovativeness of our industries depends productivity growth, which in turn has a critical bearing on the rate at which material standards of living advance." That is, a nation's success in generating innovative activity and technological advance is inextricably entwined with its ability to raise the growth of productivity, which ultimately affects international competitiveness (Audretsch and Yamawaki, 1988). As Erich Bloch emphasizes:

> In the modern marketplace, knowledge is the critical asset. It is as important a commodity as the access to natural resources or to low-skilled labor market was in the past. Knowledge has given birth to vast new industries, particularly those based on computers, semiconductors, biotechnology and designed materials.[1]

The MIT Commission on Industrial Productivity (Dertouzos et al., 1989) makes it clear that the country which is most efficient at generating and applying this knowledge will also dominate high value-added global markets.

Given this surge in interest in technological advance, surprisingly

1

little has been established in the economics literature to identify those conditions and market environments that are conducive to innovation activity and those conditions that retard it.[2] In response to this gap in scientific evidence, especially empirical, regarding the determinants of innovative activity, we have brought together a group of international scholars to present state of the art findings about innovation and technological change at a conference held at the Wissenschaftszentrum Berlin für Sozialforschung in West Berlin on August 10–11, 1989. The conference, which was jointly funded by the Wissenschaftszentrum Berlin along with the US German Marshall Fund, resulted in the set of chapters which are now included in this work.

There are five distinctive aspects emphasized throughout. The first is that most of the chapters are empirically oriented. This partly reflects the long tradition of an empirical emphasis in industrial economics at the research unit of the Wissenschaftszentrum Berlin; the empirical orientation also indicates the emergence of new sources of data in the last several years that here enabled new insights into previously untestable hypotheses.

The second feature, which follows closely from the first, is a careful consideration of measurement issues. The traditional road to operationalizing the theoretical concepts of innovation and technological change has been fraught with heroic assumptions, sometimes requiring a generous imagination. Often the results and conclusions are severely limited by the empirical measure. A number of chapters go to considerable effort to carefully explain, qualify, and contrast relatively novel measures of innovative activity.

Innovation and technological change are subjects which span the discipline of economics – for example, technical change can be analyzed from a macroeconomic as well as a microeconomic perspective. So, in keeping with the tradition of our research unit, the work also analyzes the subject through the lens of industrial economics. In particular, a number of the chapters explore the manner in which certain aspects of market structure, and in particular firm size, influence innovative activity. This is consistent with the Schumpeterian tradition, which emphasizes both market structure and firm size as being key determinants in shaping technological change.

The fourth feature reflects the long-standing tradition in industrial organization that, just as market structure and firm size are considered to influence innovative activity, technological change is viewed as having, at least in certain instances, an impact on elements related to market structure and the size distribution of firms. Several chapters explicitly analyze the impact that innovative activity has had on shaping market structure.

Finally, this work deliberately includes studies from several different countries and from both sides of the Atlantic. Not only is this in keeping with the Wissenschaftszentrum Berlin tradition of including more than one country in social science research, but this also reflects that, as for numerous commodity markets, the market for economic ideas is becoming increasingly global.

1.2 Measurement issues

A researcher trying to empirically examine the determinants and/or effects of technological change must confront an inherent challenge in measurement: virtually all empirical studies have had to rely on a proxy measure of innovative activity. Just as Kuznets (1962) observed that perhaps the greatest obstacle to understanding the role of innovation in economic processes has been the lack of meaningful measures of innovative inputs and outputs, Cohen and Levin (1989) more recently warned that:

> A fundamental problem in the study of innovation and technical change in industry is the absence of satisfactory measures of new knowledge and its contribution to technological progress. There exists no measure of innovation that permits readily interpretable cross-industry comparisons. Moreover, the value of an innovation is difficult to assess, particularly when the innovation is embodied in consumer products.

Measures of technological change have typically involved one of three aspects of the innovative process:

1. A measure of the inputs into the innovative process, such as research and development (R&D) expenditures, or else the share of the labor force accounted for by employees involved in R&D activities.
2. An intermediate output, such as the number of inventions which have been patented.
3. A direct measure of innovative output.

A clear limitation in using R&D activity as a proxy measure for innovative activity is that R&D reflects only the resources allocated toward trying to produce innovative output, but not the actual amount of resulting innovative activity. That is, R&D is an input and not an output in the innovation process. Cohen and Levin (1989) also point out that the measure of R&D employment is additionally biased because it excludes the amount of services flowing from research and laboratory materials. When combined with labor, such services often are a key input in producing innovative output. Similarly, expenditures on R&D contain an additional bias because, rather than capitalizing long-lived equipment under standard accounting rules, the expenditures are included in R&D purchases.

The reliability of patent measures has also been questioned. Not only are many innovations never patented, but not all patented inventions result in an innovation.[3] According to Scherer (1983a, pp. 107–8):

> [T]he quantity and quality of industry patenting may depend upon chance, how readily a technology lends itself to patent protection, and business decision-makers' varying perceptions of how much advantage they will derive from patent rights. Not much of a systematic nature is known about these phenomena, which can be characterized as differences in the propensity to patent.

In fact, the number of inventions which have been patented is without a doubt the most widespread proxy measure of innovative activity.[4] Still, Cohen and Levin (1989) warn, "There are significant problems with patent counts as a measure of innovation, some of which affect both within-industry and between-industry comparisons." This disparity in the propensity to patent across industries is explained by Mansfield (1984, p. 462):

> The value and cost of individual patents vary enormously within and across industries . . . . Many inventions are not patented. And in some industries, like electronics, there is considerable speculation that the patent system is being bypassed to a greater extent than in the past. Some types of technologies are more likely to be patented than others.

Thus, even as new and superior sources of patent data have been used (Hall *et al.*, 1986; Jaffe, 1986; Pakes and Griliches, 1980, 1984), the reliability of these data as measures of innovative activity has been challenged. For example, Pakes and Griliches (1980, p. 378) observe that, "patents are a flawed measure (of innovative output); particularly since not all new innovations are patented and since patents differ greatly in their economic impact".

In fact, several of the chapters included in this work use the number of patented inventions as a measure of innovative activity. However, these papers use either new sources of patent data or existing databases to compare the patent measure with other measures of innovative output. In Chapter 7, Joachim Schwalbach and Klaus Zimmermann provide one of the first applications of a rich data set containing the patent activity of West German manufacturing companies. They are able to use the files for the stock of patents held by each of 143 West German manufacturing companies in 1982 collected by the German

Patent Office.[5] By contrast, in Chapter 5 Alok K. Chakrabarti and Michael R. Halperin use a fairly standard source of data for US patents issued by the US Office of Patents and Trademarks, the BRS/PATSEARCH on-line database, to identify the number of inventions patented by over 470 firms between 1975 and 1986. Of particular interest is their comparison between the propensity of firms to patent and company R&D expenditures, and a measure not often used in industrial organization – the number of papers and publications contributed by employees of each firm. Not only do they bring together data from a number of rich sources, but they compare how the relationships between the various measures of innovative activity vary across firm size.

The two traditional methods used to measure technological change – R&D activity and patents – are among just a handful of attempts to directly measure innovative activity. The first of these sources was compiled by the Gellman Research Associates (1976) for the National Science Foundation. Gellman identified 500 major innovations that were introduced into the market between 1953 and 1973 in the United States, the United Kingdom, Japan, West Germany, France and Canada. The database was compiled by an international panel of experts, who identified those innovations representing the "most significant new industrial products and processes, in terms of their technological importance and economic and social impact" (National Science Board, 1975, p. 100).

A second and comparable database once again involved the Gellman Research Associates (1982), this time for the US Small Business Administration. In their second study, Gellman compiled a total of 635 US innovations, including 45 from the earlier study for the National Science Foundation. The additional 590 innovations were selected from fourteen industry trade journals for the period 1970–79. About 43 per cent of the sample was selected from the award winning innovations described in *Industrial Research & Development* magazine.

The third other data source that has attempted to directly measure innovation activity has been compiled at the Science Policy Research Unit (SPRU) at the University of Sussex in the United Kingdom.[6] The SPRU data consist of a survey of 4,378 innovations that were assembled over a period of fifteen years. The survey was formed by writing to experts in each industry and asking them to identify "significant technical innovations that had been successfully commercialized in the United Kingdom since 1945, and to name the firm responsible" (Pavitt *et al.*, 1987, p. 299).

The other major database providing a direct measure of innovative activity is the US Small Business Administration Innovation Data Base

(SBIDB). The database consists of 8,074 innovations introduced into the United States in 1982. A private firm, The Futures Group, compiled the database and performed quality control analyses for the US Small Business Administration by examining over 100 technology, engineering and trade journals, covering each manufacturing industry. From the sections in each trade journal listing innovations and new products, a database consisting of the innovations by four-digit standard industrial classification (SIC) was formed.[7] These data have been implemented by Acs and Audretsch (1987, 1988, 1990) to analyze the relationships between firm size and technological change, and market structure and technological change, where a direct rather than indirect measure of innovative activity is used.

In their 1990 study (Chapter 2), Acs and Audretsch compare these four databases directly measuring innovative activity and find that they generally provide similar qualitative results. For example, while the Gellman database identified small firms as contributing 2.45 times more innovations per employee than do large firms, the SBIDB finds that small firms introduce 2.38 more innovations per employee than do their larger counterparts.[8] In general these four databases reveal similar patterns with respect to the distribution of innovations across manufacturing industries and between large and small firms. These similarities emerge, despite the obviously different methods used to compile the data, especially in terms of sampling and standard of significance (Acs and Audretsch, 1990).

A comparison of one of these direct measures of innovative output – from the SBIDB – with the two more traditional measures of innovative activity – the number of patented inventions and R&D expenditures – reveals that there are distinct differences between all three measures. Table 1.1 compares the 1977 company R&D expenditures from the Federal Trade Commission's (FTC's) Line of Business survey with the total number of innovations and the total number of patented inventions between June 1976 and March 1977 (Scherer, 1983a).[9] As the simple correlation of 0.74 indicates, there is a fairly strong relationship between the ratio of patents per innovation and R&D expenditures ($millions) per innovation. While it is clear that those sectors that tend to generate substantial innovative activity also engage in relatively high patent and R&D activity, Table 1.2 shows that the correlation between these three measures is considerably less than perfect. For example, there is a striking difference in the simple correlation of 0.746 between company expenditures on R&D and the number of total innovations, and the correlation of 0.467 between the number of patented inventions and innovative activity. Also of interest is that the correlation between the output and intermediate output

Table 1.1 Comparison of innovation data with R&D and patent measures.[a]

Industry group	Total innovations	Patents	Patents/ innovation	Company R&D (millions)	R&D (millions)/ innovation
Food & tobacco	206	311	1.51	272	1.32
Textiles & apparel	29	147	5.07	65	2.24
Lumber & furniture	83	50	0.60	37	0.45
Paper	61	292	4.79	150	2.46
Chemicals (excluding drugs)	332	3 492	10.52	1 260	3.80
Drugs	170	868	5.11	449	2.64
Petroleum	24	1 046	43.58	360	15.00
Rubber & plastics	129	637	4.94	287	2.22
Stone, clay & glass	59	477	8.09	149	2.53
Primary metals	74	424	5.73	239	3.23
Fabricated metal products	340	450	1.32	246	0.72
Machinery (excluding office)	612	1 657	2.71	852	1.39
Computers & office equipment	566	1 045	1.85	1 054	1.86
Industrial electrical equipment[b]	444	836	1.88	210	0.47
Household appliances	64	232	3.63	78	1.22
Communications equipment	262	2 384	9.10	1 136	4.34
Motor vehicles & other transportation equipment[c]	152	809	5.32	1 791	11.78
Aircraft & engines	48	501	10.44	653	13.60
Guided missiles & ordnance	16	173	10.81	103	6.44
Instruments	736	1 351	1.84	652	0.89
Total[d]	4 407	17 182	3.90	10 043	2.28

Source: Acs and Audretsch (1988).

Notes: [a]Company R&D (1974) and patent (June 1976–March 1977) data are from Scherer (1983).
[b]Includes SIC 361, 362, 364 and 367.
[c]Includes SIC 371, 373, 374, 375 and 379.
[d]Includes only industries in this table.

Table 1.2 Correlation matrix of input and output measures of innovative activity.

	Total innovations	Large firm innovations	Small firm innovations	Total R&D expenditures	Company R&D expenditures
Large firm innovations	0.920	—			
Small firm innovations	0.922	0.698	—		
Total R&D expenditure	0.481	0.532	0.379	—	
Company R&D expenditures	0.746	0.737	0.672	0.764	—
Patents	0.467	0.482	0.382	0.327	0.440

Source: Acs and Audretsch (1990).

measures – innovations and patents – is greater than that between the R&D measures and patents.[10]

In fact, four of the chapters included here make use of a direct measure of innovative activity. In Chapter 2, "Changing perspectives on the firm size problem", F. M. Scherer refers to the SPRU database to illustrate some of the findings between innovative activity and firm size. Both Chapters 3 and 11 make use of the SBIDB. While we have limited our previous analyses to the aggregated four-digit SIC industry level, here the first application of these data at the firm level is provided. This enables the returns to scale from firm size and R&D inputs in generating innovative output to be explored, something that was not feasible in previous industry studies. In Chapter 11, Schulenburg and Wagner use the same data source, except aggregated to the industry level, in order to investigate the issue of simultaneity in the relationships between innovative activity, market concentration and advertising intensity. And, in Chapter 10, Felix FitzRoy and Kornelius Kraft provide a direct measure of innovative activity in the West German metalworking industries.

In his role as discussant of the Acs and Audretsch paper (Chapter 3) at the conference, Alfred Kleinknecht mentioned that one liability which the direct measure of innovative activity shares with the traditional measures of patent counts and R&D is the implicit assumption of homogeneity of units. As Cohen and Levin (1989) observe, "In most studies, process innovation is not distinguished from product innovation; basic and applied research are not distinguished from development." That is, the *market value* of the innovations, R&D expenditures and patents is generally assumed to be homogenous – an assumption which clearly contradicts real world observation.

However, several papers included in this work follow in the footsteps of Pakes (1985), Connolly *et al.* (1986) and Connolly and Hirschey (1984) in providing measures of technological change that reflect, at least to some degree, the market value of the innovative activity. In Chapter 10, Felix FitzRoy and Kornelius Kraft use 1979 data for fifty-seven West German firms in the metalworking sector, where the innovation measure is defined as the "proportion of sales consisting of products introduced within the last five years".[11] Presumably the greater the market value of a given product innovation, the higher would be the proportion of sales accounted for by new products. Similarly, Schulenburg and Wagner are able to provide one of the first applications of a measure of innovative activity in West Germany that has not received much attention outside of that country. Their measure is from the Ifo – Institut für Wirtschaftsforschung Institute and is defined as the "percentage of shipments of those products which were introduced recently into the market and are still in the entry phase".[12] Like the measure of innovative activity used by FitzRoy and Kraft, the Schulenburg and Wagner measure reflects the market value of the innovation and therefore overcomes one of the major weaknesses in most of the other direct and indirect measures of innovative activity.

Several of the studies included here are also able to go considerably beyond the usual assumption of homogeneity of R&D inputs. In Chapter 6 Kleinknecht *et al.* extend the studies (Kleinknecht, 1987 and 1989; Kleinknecht and Verspagen, 1990) examining the extent to which measured R&D represents the actual innovative inputs for large and small firms. They find that informal R&D plays an important role in complementing formal R&D in producing innovative output. Scherer also extends his 1984 study and examines the distinction between process and product R&D. Albert N. Link and John Rees, in "Firm size, university-based research and the returns to R&D", distinguish between R&D emanating from university-based associations and R&D expenditures made without the benefit of university contacts. They find that R&D expenditures made within the context of university contacts tend to generate a higher rate of return.

1.3   Firm size

"Who innovates more – the large or the small firm?" This question has generally been the essential focus of what has evolved as one of the two central tenets of the "Schumpeterian hypothesis".[13] While neoclassical economics considered the spur of competition from an industry comprised only of small firms to be uniquely suited to promote technological change, Schumpeter (1950) provided a dissenting view. The naivety of the traditional theory has been challenged by Galbraith (1956, p. 86), who argued, "There is no more pleasant fiction than that technical change is the product of the matchless ingenuity of the small man forced by competition to employ his wits to better his neighbor. Unhappily, it is a fiction."

Despite Scherer's warning (1980, p. 418) that "[t]he search for a firm size uniquely and unambiguously optimal for invention and innovation is misguided", both theoretical and empirical evidence defending each side of the debate continues to accumulate. It is clear that the bulk of R&D expenditures in the developed countries emanates from large enterprises (National Science Board, 1987). Still, Scherer (1988, pp. 4–5) summarizes the advantages that small firms may have in contributing innovations:

Smaller enterprises make their impressive contributions to innovation because of several advantages they possess compared to larger-sized corporations. One important strength is that they are less bureaucratic, without layers of "abominable no-men" who block daring ventures in a more highly structured organization. Second, and something that is often overlooked, many advances in technology accumulate upon a myriad of detailed inventions involving individual components, materials, and fabrication techniques. The sales possibilities for making such narrow, detailed advances are often too modest to interest giant corporations. An individual entrepreneur's juices will flow over a new product or process with sales prospects in the millions of dollars per year, whereas few large corporations can work up much excitement over such small fish, nor can they accommodate small ventures easily into their organizational structures. Third, it is easier to sustain a fever pitch of excitement in small organizations, where the links between challenges, staff, and potential rewards are tight. "All-nighters" through which tough technical problems are solved expediously are common.

The plethora of empirical studies relating R&D and patents to firm size is most thoroughly reviewed in Baldwin and Scott (1987) and Cohen and Levin (1989). In fact, while there are considerable ambiguities and inconsistencies in the results, it does seem that the studies relating firm's size to R&D inputs have generally reached a different conclusion to those relating firm size to the number of patented inventions. While some studies have found that expenditures on R&D increase proportionately with firm size, and others have found that they actually increase more than proportionately with firm size, there is little evidence that this relationship is anything less than proportional. Of course, different studies applying different measures of R&D for varying coverages of firm sizes have, not surprisingly, generated somewhat different results. Still, overall the empirical evidence seems to confirm Scherer's (1989, p. 2) conclusion that "[t]he relationship between size and innovative input . . . is, on average, roughly proportional." By contrast, the scant empirical evidence relating patent activity to firm size seems to suggest that the generation of patents increases at a less than proportional rate along with firm size.

In one of the most important studies, Scherer (1984) used the US Federal Trade Commission's Line of Business data to estimate the elasticity of R&D spending with respect to firm sales for 196 industries. He found evidence of increasing returns to scale (an elasticity exceeding unity) for about 20 per cent of the industries, constant returns to scale for a little less than three-quarters of the industries, and diminishing returns (an elasticity less than unity) in less than 10 per cent of the industries. These results were consistent with the findings of Soete

(1979) that R&D intensity increases along with firm size, at least for a sample of the largest US corporations.

While the Scherer and Soete studies were restricted to relatively large firms, Bound *et al.* (1984) included a much wider spectrum of firm sizes in their sample of 1,492 firms from the 1976 COMPUSTAT data. They found that R&D increases more than proportionately along with firm size for the smaller firms, but that a fairly linear relationship exists for larger firms. Despite the somewhat more ambiguous finding in still other studies (such as Comanor, 1967; Mansfield, 1968; Mansfield *et al.*, 1971), the empirical evidence seems to generally support the Schumpeterian hypothesis that research effort is positively associated with firm size.

The few studies relating patents to firm size are considerably less ambiguous. Here the findings unequivocally suggest that "the evidence leans weakly against the Schumpeterian conjecture that the largest sellers are especially fecund sources of patented inventions" (Scherer, 1982, p. 235). In one of the most important studies, Scherer (1965) used the *Fortune* annual survey of the 500 largest US industrial corporations. He related the 1955 firm sales to the number of patents in 1959 for 448 firms. Scherer found that the number of patented inventions increases less than proportionately along with firm size. Scherer's results were confirmed by Bound *et al.* (1984) in the study mentioned above. Basing their study on 2,852 companies and 4,553 patenting entities, they determined that the small firms (with less than $10 million in sales) accounted for 4.3 per cent of the sales from the entire sample, but 5.7 per cent of the patents.

In Chapter 7, Schwalbach and Zimmermann confirm these results for West Germany. Using the sample of West German manufacturing firms described in the previous section, Schwalbach and Zimmermann find that the propensity to patent is less for the largest firms than for the medium-sized firms included in their sample. Similarly, in Chapter 5 Chakrabarti and Halperin show that the largest US firms tend to generate fewer patents per million dollars of R&D than do their smaller counterparts.

How is it that R&D inputs tend to increase at least proportionately along with firm size, but the propensity to patent tends to decrease as firm size increases? One reason why this paradox has never been addressed was the lack of direct measures of innovative activity available to researchers. In utilizing the firm records of the SBIDB, Chapter 3 can relate innovative inputs to innovative outputs and provides at least one explanation for this apparent paradox. Combining individual firm records of R&D and innovative output for over 700 enterprises, we are able to determine that, although larger firms may be more R&D-

intensive than their smaller counterparts, the productivity of R&D apparently falls along with firm size (see also Acs and Audretsch's forthcoming study). That is, the empirical evidence suggests that decreasing returns to R&D expenditures in producing innovative output exist.

One limitation of Chapter 3 is that no explanations are provided as to *how* the smaller firms are able to be more productive at generating innovative output. At least some insight into this is provided in Chapter 4 by Link and Rees, who test the hypothesis that diseconomies of scale exist in generating innovative activity because "bureaucratization in the innovation decision-making process inhibits not only inventiveness but also slows the pace at which new inventions move through the corporate system toward market." They provide evidence suggesting that large and small firms have different contacts with universities, and that small firms appear to be more efficient at exploiting university-based associations, thereby enabling them to raise the productivity of their internal R&D.

By contrast, the study by Scherer (Chapter 2) suggests that small firms may tend to have a relative advantage at product innovation, while their larger counterparts may have a relative advantage at contributing process innovations. In fact, Scherer finds considerable evidence supporting the hypothesis that large firms tend to be more process R&D-oriented than are small firms. Similarly, in Chapter 12 Wesley M. Cohen and Steven Klepper provide a basis for why industries composed of many small firms will tend to exhibit greater diversity in innovative activity as well as more rapid technological change.

1.4   Market structure

In comparison to the number of studies investigating the relationship between firm size and technological change, those examining the market structure/technological change relationship are what Baldwin and Scott (1987, p. 89) term "miniscule" in number. In fact, the most comprehensive and insightful evidence has been made possible by utilizing the Federal Trade Comission's Line of Business data. Using 236 manufacturing industry categories, which are defined at both the three and four-digit SIC level, Scherer (1983b) found that 1974 company R&D expenditures divided by sales were positively related to the 1974 four-firm concentration ratio. Scherer (1983b, p. 225) concluded that, "although one cannot be certain, it appears that the advantages a high market share confers in appropriating R&D benefits provide the most likely explanation of the observed R&D-concentration associations."

Scott (1984) also used the FTC Line of Business survey data and found the U-shaped relationship between market concentration and R&D that Scherer (1980) had earlier described. However, when he controlled for the fixed effects for two-digit SIC industries, no significant relationship could be found between concentration and R&D. These results are consistent with a series of studies by Levin *et al.* (1985, 1987), Levin and Reiss (1984) and Cohen *et al.* (1987). Using data from a survey of R&D executives in 130 industries, which were matched with FTC Line of Business industry groups, Cohen *et al.* (1987) and Levin *et al.* (1987) found little support for the contention that industrial concentration is a significant and systematic determinant of R&D effort.

Although these studies have produced somewhat ambiguous results, finding either a positive relationship between concentration and R&D or else no significant relationship between concentration and R&D, a negative relationship is usually not identified. Thus, there might exist some uncertainty about the exact nature of the relationship, but it appears unlikely that lower levels of concentration are associated with greater R&D effort.

However, when we (1988, 1990) relate market concentration to their direct measure of innovative activity, a different relationship emerges. Unequivocal evidence is found that concentration exerts a negative influence on the number of innovations being made in an industry.

While it has been hypothesized that firms in concentrated industries are better able to capture the rents accruing from an innovation, and therefore have a greater incentive to undertake innovative activity, there are other market structure variables that also influence the ease with which economic rents can be appropriated. For example, Comanor (1967) argued and found that, based on a measure of minimum efficient scale, there is less R&D effort (average number of research personnel divided by total employment) in industries with very low scale economies. However, he also found that in industries with a high minimum efficient scale, R&D effort was also relatively low. Comanor interpreted his results to suggest that where entry barriers are relatively low, there is little incentive to innovate, since the entry subsequent to an innovation would quickly erode any economic rents. At the same time, in industries with high entry barriers, the absence of potential entry may reduce the incentive to innovate.

Using their sample of West German manufacturing firms, Schwalbach and Zimmermann (Chapter 7) find that the amount of patents registered (by both parent companies and subsidiaries) is positively related to advertising intensity, but negatively related to the Herfindahl Index. Even after controlling for the technological opportunity class, Schwalbach and Zimmermann find that there is less and not more patent activity as the degree of market concentration increases. While Link and Rees (Chapter 4) find that the probability of a firm engaging in a university-based research relationship increases with enterprise size, there is no evidence suggesting that the university–firm research link is influenced by industry concentration. As they emphasize, this is an issue that needs to be more thoroughly researched in the future.

In Chapter 9, John Scott uses the FTC Line of Business data to test the hypothesis that the diversity of R&D projects undertaken in an industry is promoted by the extent of market competition. Using a novel measure of research diversity, Scott shows that rivalry among firms tends to lead to the pursuit of unique R&D strategies.

1.5   The influence of innovation on market structure

There is a long-standing tradition in the field of industrial organization that, just as market structure and firm size are considered to influence innovative activity, technological change is viewed as having, at least in certain instances, an impact on market structure and the size distribution of firms (Dasgupta, 1986; Stiglitz and Mathewson, 1986; and Dasgupta and Stiglitz, 1980). As early as 1948 Blair observed that,

> The whole subject of the comparative efficiency of different sizes of business has long raised one of the most perplexing dilemmas in the entire body of economic theory . . . . But a beginning must be made sometime in tackling this whole size-efficiency problem on an empirical basis. The first step in any such undertaking would logically be that of studying the underlying technological forces of the economy, since it is technology which largely determines the relationship between the size of plant and efficiency. (1948, p. 121)

In fact, Blair argued that due to particular innovations and fundamental shifts in technology, the trend toward increasing plant size and market concentration had been replaced by the opposite trend toward smaller size and less industrial concentration.

In 1966 Phillips suggested that market structure may be endogenous to technological change. He provided the example of the civilian aircraft industry as an illustration of the manner in which innovative activity can shape market structure (Phillips, 1971). While Levin (1978) argued that both market concentration and changes in scale economies are endogenous, Nelson and Winter(1978, 1982) provided simulation models demonstrating the manner in which innovation determines subsequent market structure. Most recently, Dosi (1988) emphasized the importance of examining the dynamic process by which technical change and market structure interact with each other in determining the direction of market evolution.

Mansfield (1962) provided one of the first empirical studies relating innovative activity to subsequent firm performance. He found that firm growth tends to be greater, *ceteris paribus*, following an innovation and that innovative activity can influence the degree of market concentration. Building on the theoretical arguments of Nelson and Winter (1982), Winter (1984), Gort and Klepper (1982), and Klepper and Graddy (1989), we find (Acs and Audretsch, 1989a, b) that the entry of new firms in an industry is influenced by the technological regime – the entrepreneurial regime tends to promote the entry of new firms, while firms are deterred from entering industries characterized by the technological regime.

Carlsson (1989) and Acs *et al.* (1990) show that the implementation of flexible technology has led to a striking decrease in both firm and plant size in the metalworking industries. That is, the diffusion of new technology has resulted in a pronounced effect on the size distribution of firms.

Two chapters explicitly address the issue of simultaneity between technological change and market structures and the impact of innovation on the firm growth. In "Advertising, innovation and market structure: a comparison of the United States of America and the Federal Republic of Germany", Schulenburg and Wagner consider the existence of a simultaneous relationship between innovative activity and two crucial aspects of market structure – concentration and advertising.[14] Comparing a simultaneous system model with an ordinary least squares model, for both the United States and West Germany, Schulenburg and Wagner find that, in fact, both of these measures of market structure and innovation are endogenously determined. The consistency in results between the two countries suggests that these relationships are fairly robust.

In Chapter 10, FitzRoy and Kraft use the sample of West German manufacturing firms previously described to test the hypothesis raised by Mansfield (1962) that innovative activity promotes subsequent firm growth. Even after controlling for firm age, FitzRoy and Kraft find that firm sales rates are positively influenced by their extent of innovative activity as well as the degree of skilled labor in the labor force.

1.6   Conclusions

As an increasing amount of public attention has focused on the subject of technological change, there has also been a corresponding rush of policy proposals designed to promote innovative activity and ultimately international competitiveness and productivity. Many of these proposals, however, are apparently oblivious to what is currently known about technological change and innovation in the industrial organization literature. It is ironic that, just when systematic evidence is beginning to accumulate that the contributions to innovative activity from small firms are at least as important as those from larger firms, public policy seems to be moving in the opposite direction. For example, in 1986 the Hon. Malcolm Baldridge, Secretary of Commerce, asserted, "We are simply living in a different world today. Because of larger markets, the cost of research and development, new product innovation, marketing, and so forth . . . it takes larger companies to compete successfully" (Baldridge, 1986). Baldridge pointed out that the American share of the largest corporations in the world had fallen considerably between 1960 and 1984. He warned that programs promoting large scale enterprise must "not be stopped by those who are preoccupied with outdated notions about firm size."[15]

An example of the conventional wisdom regarding firm size and innovative activity surfaced in the Reagan Administration's proposed changes in the antitrust laws, particularly in the areas of merger policy, joint ventures and cooperative R&D arrangements.[16] This conventional wisdom is also evident in the industrial policies emanating from the Commission of the European Communities in directing the changes for the completion of the European internal market in 1992.[17]

In fact, as is clear from a number of the contributions included here, there is substantial evidence that, contrary to this line of conventional wisdom, both small and large firms play an important role in contributing to innovative activity. Further, these results appear to hold not just for US firms, but, as several of the chapters find, also for European firms.

A number of the chapters also make it clear that, in order to better understand the relationships between firm size, market structure and innovative activity, new and alternative measures of technological change must be developed. The earliest empirical studies addressing the impact of firm size on technological change focused on an input measure of R&D expenditures or employees. With the advent of patents as a measure of technological change came a measure that was not only an intermediate output, rather than an input, but also seemed to better capture innovative efforts across a greater spectrum of firm sizes. In this work, technological change is alternatively measured by the share of sales accounted for new products, the number of innovations contributed, informal as well as formal R&D, product as well as process R&D, and the effects of university contacts. Future research is likely to include a greater variety of measures of technological change, thereby enabling a richer identification of those market conditions and firm characteristics that are most conducive to innovative activity.

Notes

1. Erich Bloch, "Can the US compete?", *World Link*, January/February 1990, p. 9.
2. For an excellent review of the state of knowledge about the determinants of innovation and technological change, see Baldwin and Scott (1987) and Cohen and Levin (1989).
3. The distinction between an innovation and an invention is made clear by Edwards and Gordon (1984, p. 1), who define an innovation as "a process that begins with an invention, proceeds with the development of the invention, and results in the introduction of a new product, process or service to the marketplace."
4. According to Shepherd (1979, p. 40), "Patents are a notoriously weak measure. Most of the eighty thousand patents issued each year are worthless and are never used. Many are of moderate value, and a few are bonanzas. Still others have negative social value. They are used as 'blocking' patents to stop innovation, or they simply are developed to keep competition out."
5. For a more detailed explanation of the patent data compiled by the West German Patent Office (Deutsches Patentamt), see Greif (1989) and Bundesminister für Forschung und Technologie (1982).
6. The SPRU innovation data are explained in more detail in Pavitt *et al.* (1987), Townsend *et al.* (1981), Robson and Townsend (1984) and Rothwell (1989).
7. A detailed description of the US Small Business Administration Innovation Data Base can be found in Chapter 2 of Acs and Audretsch (1990).
8. See Scherer (1989) for further explanations.
9. Patents represent 59 per cent of all patented inventions issued to US corporations between June 1976 and March 1977 and 61 per cent of patents issued to industrial corporations over the same period (Scherer, 1983).
10. For further analysis comparing the innovation and patent measures see Acs and Audretsch (1989c).
11. For another application of these data see FitzRoy and Kraft (1990).
12. The database used by Graf von der Schulenburg and Wagner is the *Ifo Innovations Test* and is explained in greater detail in Oppenländer (1990), and König and Zimmermann (1986).
13. Kamien and Schwartz (1975, p. 15) characterize the Schumpeterian debate thus: "A statistical relationship between firm size and innovative activity is most frequently sought with exploration of the impact of firm size on both the amount of innovational effort and innovation success."
14. A similar examination of simultaneity between innovative activity, market structure and unionization can be found in Audretsch and Schulenburg (1990).

15. Statement of the Hon. Malcolm Baldridge, Secretary, Department of Commerce, in *Merger Law Reform: hearing on S. 2022 and S. 2160 before the Senate Committee on the Judiciary*, 99th Congress, 2nd Session, 1986.
16. For an example of these proposed changes in the antitrust laws, see *Antitrust Policy and Competition: Hearings before the Joint Economic Committee*, 98th Congress, 2nd Session, 1986.
17. For a brief discussion of how the conventional wisdom regarding firm size and innovative activity is shaping the industrial policy of the Commission of the European Communities, see David B. Audretsch, "America's challenge to Europe", *The Wall Street Journal*, July 31, 1989, p. 6.

References

Acs, Zoltan J. and David B. Audretsch, *Innovation and Small Firms*, Cambridge, MA: MIT Press, 1990.

Acs, Zoltan J. and David B. Audretsch, "Small-firm entry in US manufacturing", *Economica*, vol. 56, May 1989a, pp. 255–65.

Acs, Zoltan J. and David B. Audretsch, "Births and firm size", *Southern Economic Journal*, vol. 56, October 1989b, pp. 467–75.

Acs, Zoltan J. and David B. Audretsch, "Patents as a measure of innovative activity", *Kyklos*, vol. 42, 1989c, pp. 171–80.

Acs, Zoltan J. and David B. Audretsch, "Innovation in large and small firms: An empirical analysis", *American Economic Review*, vol. 78, September 1988, pp. 678–90.

Acs, Zoltan J. and David B. Audretsch, "Innovation, market structure and firm size", *Review of Economics and Statistics*, vol. 69, November 1987, pp. 567–75.

Acs, Zoltan J., David B. Audretsch and Bo Carlsson, "Flexibility, plant size, and industrial restructuring", in Zoltan J. Acs and David B. Audretsch, eds., *The Economics of Small Firms: A European Challenge*, Boston: Kluwer Academic Publishers, 1990, pp. 141–54.

Audretsch, David B. and Zoltan J. Acs, "Innovation and size at the firm level", *Southern Economic Journal*, forthcoming (1991).

Audretsch, David B. and J.-Matthias Graf von der Schulenburg, "Union participation, innovation, and concentration: Results from a simultaneous model", *Journal of Institutional and Theoretical Economics*, vol. 146, 1990, pp. 298–313.

Audretsch, David B. and Hideki Yamawaki, "R&D rivalry, industrial policy and US–Japanese trade", *Review of Economics and Statistics*, vol. 70, August 1988, pp. 438–47.

Baldridge, Malcolm, "The Administration's legislative proposal and its ramifications", *Antitrust Law Journal*, vol. 29, 1986.

Baldwin, William L. and John T. Scott, *Market Structure and Technological Change*, London and New York: Harwood Academic Publishers, 1987.

Blair, John M., "Technology and size", *American Economic Review*, vol. 38, May 1948, pp. 121–52.

Bound, John, Clint Cummins, Zvi Griliches, Bronwyn H. Hall and Adam Jaffe, "Who does R&D and who patents?" in Zvi Griliches, ed., *R&D, Patents and Productivity*, Chicago, IL: University of Chicago, 1984, pp. 21–54.

Bundesminister für Forschung und Technologie, *Die Messung wissenschaftlicher und technischer Tätigkeiten*, Bonn: Bundesminister für Forschung und Technologie, 1982.

Carlsson, Bo, "The evolution of manufacturing technology and its impact on industrial structure: An international study", *Small Business Economics*, vol. 1, January 1989, pp. 21–38.

Cohen, Wesley M. and Richard C. Levin, "Empirical studies of innovation and market structure", in Richard Schmalensee and Robert Willig, eds., *Handbook of Industrial Organization*, vol. II, Amsterdam: North-Holland, 1989.

Cohen, Wesley M., Richard C. Levin and David C. Mowery, "Firm size and R&D intensity: A reexamination", *Journal of Industrial Economics*, vol. 35, June 1987, pp. 543–65.

Comanor, William S., "Market structure, product differentiation and industrial research", *Quarterly Journal of Economics*, vol. 81, November 1967, pp. 639–57.

Connolly, Robert A. and Mark Hirschey, "R&D, market structure and profits: A value based approach", *Review of Economics and Statistics*, vol. 66, November 1984, pp. 682–6.

Connolly, Robert A., Barry T. Hirsch and Mark Hirschey, "Union rent seeking, intangible capital and the market value of the firm", *Review of Economics and Statistics*, vol. 68, November 1986, pp. 567–77.

Dasgupta, Partha, "The theory of technological competition", in J.E. Stiglitz and G.F. Mathewson, eds., *New Developments in the Analysis of Market Structure*, Cambridge, MA: MIT Press, 1986.

Dasgupta, Partha and Joseph Stiglitz, "Industrial structure and the nature of innovative activity", *Economic Journal*, vol. 90, June 1980, pp. 266–93.

Dertouzos, Michael L., Richard K. Lester, Robert M. Solow and the MIT Commission on Industrial Productivity, *Made in America: Regaining the Productive Edge*, Cambridge, MA: MIT Press, 1989.

Dosi, Giovanni, "Sources, procedures and microeconomic effects of innovation", *Journal of Economic Literature*, vol. 26, September 1988, pp. 1120–71.

Edwards, Keith L. and Theodore J. Gordon, "Characterization of innovations introduced on the US market in 1982", The Futures Group, prepared for the US Small Business Administration under contract no. SBA-6050-0A-82, March 1984.

FitzRoy, Felix R. and Kornelius Kraft, "Innovation, rent-sharing and the organization of labour in the Federal Republic of Germany", *Small Business Economics*, vol. 2, 1990, pp. 95–104.

Galbraith, John K., *American Capitalism: The Concept of Countervailing Power*, revised edition, Boston, MA: Houghton Mifflin, 1956.

Gellman Research Associates, "The relationship between industrial concentration, firm size and technological innovation", prepared for the Office of Advocacy, US Small Business Administration under award no. SBA-2633-0A-79, May 1982.

Gellman Research Associates, "Indicators of international trends in technological innovation", prepared for the National Science Foundation, April 1976.

Gort, Michael and Steven Klepper, "Time paths in the diffusion of product innovations", *Economic Journal*, vol. 92, September 1982, pp. 630–53.

Greif, Siegfried, "Zur Erfassung von Forschungs- und Entwicklungstätigkeit durch Patente", *Naturwissenschaften*, vol. 76, April 1989, pp. 156–9.

Hall, Bronwyn H., Zvi Griliches and Jerry A. Hausman, "Patents and R&D: Is there a lag?" *International Economic Review*, vol. 27, June 1986, pp. 265–302.

Jaffe, Adam B., "Technological opportunity and spillovers of R&D: Evidence from firms' patents, profits and market value", *American Economic Review*, vol. 76, December 1986, pp. 984–1001.

Kamien, Morton I. and Nancy L. Schwartz, "Market structure and innovation: A survey", *Journal of Economic Literature*, vol. 13, March 1975, pp. 1–37.

Kleinknecht, Alfred, "Firm size and innovation: Observations in Dutch manufacturing industry", *Small Business Economics*, vol. 1, 1989, pp. 215–22.

Kleinknecht, Alfred, "Measuring R&D in small firms: How much are we missing?" *Journal of Industrial Economics*, vol. 34, 1987, pp. 253–6.

Kleinknecht, Alfred and Bart Verspagen, "R&D and market structure: The impact of measurement and aggregation problems", *Small Business Economics*, vol. 1, 1989, pp. 297–302.

Klepper, Steven and Elizabeth Graddy, "The evolution of new industries and the determinants of market structure", unpublished manuscript, August 1989.

König, Heinz and Klaus F. Zimmermann, "Innovations, market structure and market dynamics", *Journal of Institutional and Theoretical Economics*, vol. 142, 1986, pp. 184–99.

Kuznets, Simon, "Inventive activity: Problems of definition and measurement", in R.R. Nelson, ed., *The Rate and Direction of Inventive Activity*, Princeton, NJ: Princeton University Press, 1962, pp. 19–43.

Levin, Richard C., "Technical change, barriers to entry and market structure", *Economica*, vol. 45, 1978, pp. 347–61.

Levin, Richard C., Wesley Cohen and David C. Mowery, "R&D appropriability, opportunity and market structure: New evidence on the Schumpeterian hypothesis", *American Economic Review*, vol. 15, May 1985, pp. 20–4.

Levin, Richard C., Alvin K. Klevorick, Richard R. Nelson and Sydney G. Winter, "Appropriating the returns from industrial research and development", *Brookings Papers on Economic Activity*, vol. 3, 1987, pp. 783–820.

Levin, Richard C. and Peter C. Reiss, "Tests of a Schumpeterian model of R&D and market structure", in Zvi Griliches, ed., *R&D, Patents and Productivity*, Chicago, IL: University of Chicago, 1984, pp. 175–208.

Mansfield, Edwin, "Comment on using linked patent and R&D data to measure interindustry technology flows", in Zvi Griliches, ed., *R&D, Patents and Productivity*, Chicago, IL: University of Chicago, 1984, pp. 462–4.

Mansfield, Edwin, *Industrial Research and Technological Change*, New York, NY: W. W. Norton for the Cowles Foundation for Research Economics, Yale University, 1968, pp. 83–108.

Mansfield, Edwin, "Entry, Gibrat's law, innovation and the growth of firms", *American Economic Review*, vol. 52, December 1962, pp. 1023–51.

Mansfield, Edwin, J. Rapoport, J. Schnee, J. Wager and M. Hamburger, *Research and Innovation in the Modern Corporation*, New York: W. W. Norton, 1971.

National Science Board, *Science and Engineering Indicators – 1987*, Washington,

DC: US Government Printing Office, 1987.

National Science Board, *Science Indicators – 1974*, Washington, DC: US Government Printing Office, 1975.

Nelson, Richard R. and Sidney G. Winter, *An Evolutionary Theory of Economic Change*, Cambridge, MA: Harvard University Press, 1982.

Nelson, Richard R. and Sidney G. Winter, "Forces generating and limiting concentration under Schumpeterian competition", *Bell Journal of Economics*, vol. 9, Autumn 1978, pp. 534–48.

Oppenländer, Karl Heinz, "Investitionsverhalten und Marktstruktur – empirische Ergebnisse für die Bundesrepublik Deutschland", in B. Gahlen, ed., *Marktstruktur und gesamtwirtschaftliche Entwicklung*, Berlin: Springer-Verlag, 1990, pp. 253–66.

Pakes, Ariel, "On patents, R&D and the stock market rate of return", *Journal of Political Economy*, vol. 93, April 1985, pp. 390–409.

Pakes, Ariel and Zvi Griliches, "Patents and R&D at the firm level: A first look," in Zvi Griliches, ed., *R&D, Patents and Productivity*, Chicago, IL: University of Chicago, 1984, pp. 55–72.

Pakes, Ariel and Zvi Griliches, "Patents and R&D at the firm level: A first report", *Economics Letters*, vol. 5, 1980, pp. 377–81.

Pavitt, Keith, Michael Robson and Joe Townsend, "The size distribution of innovating firms in the UK: 1945–1983", *Journal of Industrial Economics*, vol. 55, March 1987, pp. 291–316.

Phillips, Almarin, *Technology and Market Structure*, Lexington, MA: D. C. Heath, 1971.

Phillips, Almarin, "Patents, potential competition and technical progress", *American Economic Review*, vol. 56, 1966, pp. 301–10.

Robson, Michael and Joe Townsend, "Users manual for ESRC archive file on innovations in Britain since 1945: 1984 update", Brighton: Science Policy Research Unit, University of Sussex, 1984.

Rothwell, Roy, "Small firms, innovation and industrial change", *Small Business Economics*, vol. 1, January 1989, pp. 51–64.

Santarelli, Enrico and Alessandro Sterlachinni, "Innovation, formal vs. informal R&D and firm size: Some evidence from Italian manufacturing firms", *Small Business Economics*, vol. 2, 1990, pp. 223–8.

Scheirer, William K., "SBA innovation database: A comment", *Small Business Economics*, vol. 1, 1989, p. 231.

Scherer, F. M., "Does antitrust compromise technological efficiency?", *Eastern Economic Journal*, vol. 15, January–March 1989, pp. 1–5.

Scherer, F. M., "Innovation and small firms", Testimony before the Subcommittee on Monopolies and Commercial Law, Committee on the Judiciary, US House of Representatives, February 24, 1988.

Scherer, F. M., *Innovation and Growth: Schumpeterian Perspectives*, Cambridge, MA: MIT Press, 1984.

Scherer, F. M., "The propensity to patent", *International Journal of Industrial Organization*, vol. 1, March 1983a, pp. 107–28.

Scherer, F. M., "Concentration, R&D, and productivity change", *Southern Economic Journal*, vol. 50, July 1983b, pp. 221–5.

Scherer, F. M., "Inter-industry technology flows in the United States", *Research Policy*, vol. 11, 1982, pp. 227–45.

Scherer, F. M., *Industrial Market Structure and Economic Performance*, 2nd edition, Chicago, IL: Rand McNally College Publishing Co., 1980.

Scherer, F. M., "Firm size, market structure, opportunity and the output of patented inventions", *American Economic Review*, vol. 55, December 1965, pp. 1097–125.

Schumpeter, Joseph A., *Capitalism, Socialism and Democracy*, 3rd edition, New York, NY: Harper and Row, 1950.

Scott, John T., "Firm versus industry variability in R&D intensity", in Zvi Griliches, ed., *R&D, Patents and Productivity*, Chicago, IL: University of Chicago Press, 1984, pp. 233–48.

Shepherd, William G., *The Economies of Industrial Organization*, Englewood Cliffs, NJ: Prentice Hall, 1979.

Soete, Luc L. G., "Firm size and inventive activity: The evidence reconsidered", *European Economic Review*, vol. 12, 1979, pp. 319–40.

Stiglitz, Joseph E. and G. Frank Mathewson, eds., *New Developments in the Analysis of Market Structure*, Cambridge, MA: MIT Press, 1986.

Townsend, Joe, F. Herwood, G. Thomas, Keith Pavitt and S. Wyatt, "Innovations in Britain since 1945", Occasional Paper No. 16, Brighton: Science Research Unit, University of Sussex, 1981.

Winter, Sidney G., "Schumpeterian competition in alternative technological regimes", *Journal of Economic Behavior and Organization*, vol. 5, September–December 1984, pp. 287–320.

2

Changing perspectives on the firm size problem

F. M. Scherer

2.1   A voyage into the past

The invitation to speak in Berlin on technological change and firm size launched me on a kind of sentimental journey. Twenty-four years ago I was similarly invited to participate in a conference on "Research and Development, the Trend of Technological Change, and its International Consequences" at the Institut Européen d'Administration des Affaires (INSEAD) in Fontainebleau. It was for me an exciting venture – my first trip to Europe for professional purposes other than demonstrating prowess as a soldier. And at the time, I was well primed for the occasion, since I had been working for eight years on various aspects of the economics of technology and had just completed a major research project on the links between firm size, market structure and the vigor of technological innovation in the United States.

That project had been inspired by J. A. Schumpeter's argument that the large, monopolistic corporation provided an ideal environment for advancing technology. Some previous research by economists had raised doubts about the proposition, but their data and methodology left much to be desired, and I hoped to clarify matters by marshalling superior data and using more sophisticated statistical tests. As the chips fell, the evidence went strongly against Schumpeter, leading me to conclude:

> The principal conclusions of this study are as follows: (1) Inventive output increases with firm sales, but generally at a less than proportional rate. (2) Differences in technological opportunity . . . are a major factor responsible for interindustry differences in inventive output. (3) Inventive output does

not appear to be systematically related to variations in market power, prior profitability, liquidity, or (when participation in fields with high technological opportunity is accounted for) degree of product line diversification.

These findings among other things raise doubts whether the big, monopolistic, conglomerate corporation is as efficient an engine of technological change as disciples of Schumpeter (including myself) have supposed it to be. Perhaps a bevy of fact-mechanics can still rescue the Schumpeterian engine from disgrace, but at present the outlook seems pessimistic.[1]

There is no fervor like that of a recent convert, and so for me, the Fontainebleau conference was a shock akin to what a newly confirmed atheist might experience if suddenly thrust into the front row at a fundamentalist revival. At the time, Europe, and especially France, was firm-size mad. In the wisdom of hindsight, I understand that the conference had been convened to provide support for the French Government's fifth plan, whose objective was "the constitution, or the reinforcement where extant, of a small number of firms or groups of firms, international in dimension, which are capable of confronting foreign groups."[2] The keynote speaker, Finance Minister Valéry Giscard d'Estaing, observed that the United States was home to most of the world's largest industrial firms and wondered whether this might be why research and development outlays per capita were so much higher in the United States than in Europe. Others hammered at the same theme. A senior INSEAD professor stated that, because of their size, US companies were more competitive, able to proceed more quickly in the tasks of R&D and better able to amortize the costs of R&D. A well-known American scholar pointed out that the United States was home to 248 companies with sales exceeding $250 million, while France had only 21, Great Britain 49 and Germany 27 enterprises of that size. He inferred from this that large US corporations could better withstand the risks of R&D, could afford better process technology, could absorb larger marketing and production costs, and benefited from being able to assemble large vertically integrated systems. General Motors and the United States Steel Corporation, each with outputs exceeding that of whole European national industries, were cited as examples of particularly impressive technological power. An official of France's Delegation Générale à la Recherche Scientifique et Technique recommended that for European firms to regain techological parity with their American counterparts, the Europeans had to redouble their research efforts; R&D expenses needed to be internationally shared; the growth of American companies could be retarded by the enactment of international antitrust laws; and American firms should be required to establish laboratories within Europe and to license their technology to European companies. The final plenary session speaker recorded in my notes, an official of the French Finance Ministry, concluded that state intervention was needed to foster technological progress in fields where the pace of advance was most rapid and also in traditional industries with productivity enhancement prospects.

To be sure, there were dissident voices. Another senior American scholar urged that in addition to discussing the advantages of big business, the conference might ask why small businesses did as well in advancing technology as they seemed to do. An IBM USA research director suggested that although firm size was a positive factor in the development of ideas, it was a negative factor in the generation of ideas. A Swiss research director had the temerity to propose that the differences between the United States and Europe were more a matter of motivation than industrial organization, and in particular, that in Europe, aggressive, competitive attitudes toward pushing new technology were absent. But these were voices crying in the wilderness; the conference was predestined to conclude that building giant enterprises was how Europe could cope with the American technological challenge.[3]

2.2 International differences and their reasons

Persons thrust into an alien culture professing unfamiliar beliefs are well advised not to conclude that they are dealing with primitives. Rather, they should question whether there might be objective circumstances that make what is false in one environment true in another. That there may have been such differences is suggested by research at the University of Sussex Science Policy Research Unit. During the period from 1956 through 1965, UK manufacturing firms with fewer than 500 employees originated significant innovations at a rate only 83 per cent of their share of total employment, according to my summary weighting of more detailed statistics published by Pavitt *et al.*[4] Companies with 50,000 or more employees on the other hand originated innovations at a rate 1.95 times their employment share. Thus, small firms were under-performers, while large firms were stars.[5] By 1971–83, however, the situation had changed. Companies with fewer than 500 employees contributed innovations at an average rate 1.39 times their employment share, while the 50,000-plus giants achieved a 1.70 rate. Both size categories were over-performers; conversely, the *least* progressive firms in relation to their size were those with employment between 5,000 and 9,999. Evidently, something changed to make the UK innovation size distribution more like what has been repeatedly observed for the United States, although the contributions of the very largest companies continue to be disproportionate in the United Kingdom, but less than proportionate, or at best proportionate, in the United States.[6]

For France, it is more difficult to reconcile the disparity of views with objective conditions. An analysis of French data for the same time period as the Fontainebleau conference revealed that small firms were if anything disproportionate supporters of research and development.[7] Similarly, a tally of important 1953–75 innovations shows France to have had the *largest* fraction of inventions, 57 per cent, originating in relatively small companies, i.e. those with sales of less than $50 million.[8] The United States was second with 50 per cent, West Germany third at 37 per cent, the United Kingdom fourth at 33 per cent and Japan last at 20 per cent.

The weak innovative record of small firms in Japan appears fully consistent with known features of the Japanese economy. The best science and engineering graduates of Japanese universities traditionally aspire to long-term positions with large, well-known corporations, which somehow appear to have maintained an unusually open and receptive environment for technological innovation.[9] Small business on the other hand is the refuge of less well-trained individuals and those who have received a "golden handshake" at the early age of 55.

It should not be surprising, especially to one who received his graduate education within the ambit of Boston's Route 128 high-technology region, that the United States has been a particularly fertile environment for small firm innovative contributions – quite different from that of Japan and perhaps also some European nations. David Mowery's painstaking historical research reveals that relatively small companies held their own in innovative activities through much of the twentieth century.[10] But a decisive change may have materialized with the emergence of the high technology venture capital industry, led by the formation in 1946 of the Boston-based American Research and Development Corporation (ARDC).[11] The Boston climate spawned new high technology enterprises founded both by academicians – by one count, 200 such firms as of 1966 from MIT laboratories and departments alone[12] – and technical staff from larger corporations such as Raytheon and Sylvania.[13] From Boston, the high technology venture movement spread west and south, most notably to Silicon Valley. Although there are indications of its further and more recent spread to the United Kingdom,[14] high technology venturing remains the United States' powerful but not-so-secret weapon in international technological competition.

2.3   Mounting skepticism

This, at least, has been the conventional wisdom among those who study the dynamics of research and innovation in the United States. Recently, however, doubts have arisen as to whether the United States had in fact chosen the right mix of innovation-fostering industrial organization policies. Reconsideration of old views has been forced by disappointing productivity growth, increasing technological competition from other nations, especially Japan, and the loss of American leadership in many technologies, e.g. consumer electronics, automobile tires, random access memory chips, machine tools, industrial enzymes, telephone switches, steel processing, facsimile devices, automobiles and even high performance aircraft structural composites. *Le défi Japonais* has shocked Americans into a scrutiny of past errors and new solutions similar to that which *Le défi Américain* stimulated among Europeans during the 1960s.

Indeed, in some respects the parallels are striking. In March 1986, there was testimony on proposed merger reform by US Secretary of Commerce Malcolm Baldridge astoundingly like the arguments heard at the 1966 Fontainebleau conference.[15] Like a speaker at Fontainebleau, but with the hemispheres reversed, Mr Baldridge observed with dismay that nineteen of the world's thirty largest corporations were foreign-owned. American companies, he continued, had to become larger, among other things, by relaxing the merger laws' constraint on their "ability . . . to combine and restructure domestically to meet new competition from abroad."

In 1984, the US Congress passed a National Cooperative Research Act establishing procedures under which the antitrust implications of inter-firm R&D joint ventures were to be judged on a rule-of-reason basis, and subjecting any antitrust violations by approved R&D joint ventures to liability for at most single damages. The law's passage reflected a general retrenchment of antitrust policy under the Reagan Administration and the perception that since joint R&D ventures were being used with evident success by overseas competitors, US firms should not be denied equivalent opportunities. There is little evidence on the extent to which important technical advances have been achieved under the new law. We do know, however, that the first batch of registered joint ventures has tended to involve relatively large firms in industries of above-average concentration and research intensity – attributes not traditionally associated with free-market incentive failures.[16] In 1989 the US Congress was considering legislation (H.R. 1,024) that would extend antitrust exemptions to the marketing and production phases of joint innovative ventures. According to a sponsor:

The economic realities behind this piece of legislation seem clear enough: First, the costs of developing many items on the new technological agenda and bringing them to the marketplace often exceed the resources of a single firm. Second, the economics of high-technology manufacturing has shifted to short-period life-cycles and the need for rapid response to changing consumer demands. Third, because newly developed products may now give a technology innovator no more than 6 months to a year of exclusivity, American firms must develop and initiate marketing strategies and production scheduling with much greater dispatch than in the past. Fourth – and equally important to all of the above – American-style competition can coexist with cooperative efforts to enhance innovation and be the spur to greater manufacturing success as in the past.[17]

How a group of firms organize themselves for joint work and secure the necessary legal approvals without consuming six months (or much more) of technological lead time was not made clear in the speaker's analysis. The experiences of the most important production joint ventures, NUMMI and SEMATECH, are hardly encouraging.

The "et Tu Brute" of recent revisionist thought was a much debated article by Charles Ferguson in the *Harvard Business Review* arguing that, for all their merits, high technology ventures are an important *cause* of America's technological competitiveness problems.[18] Two main points were stressed:

(a) that large-scale organizations are needed to perfect innovations and bring their costs down to levels competitive in world markets; and

(b) that the very ease of forming new high technology ventures weakens the large corporations that would otherwise perform those crucial functions, depleting their ranks of top scientists and engineers and, because of high turnover, preventing them from supporting long-term R&D and the training of skilled operatives.

The result, argues IBM alumnus Ferguson, is a fragmented "chronically entrepreneurial" industry unable to compete with the stable Japanese electronics giants.

2.4 A reassessment of the arguments

However contrary they may be to several variants of our conventional wisdom, these are serious charges that deserve to be taken seriously. A reassessment of the evidence seems necessary to determine whether they might be correct in part or *in toto*.

The costs of R&D

It has always been recognized that some innovative challenges are so massive that they can only be undertaken by large organizations, and perhaps then only with the help of government subsidies. The most prominent cases are major weapon systems development programs. For good reasons or bad, the costs of weapons programs have been escalating at an impressive rate. The all-time record was set recently by the B-2 (Stealth) bomber program, on which more than $22 billion was spent before the first prototype flight in 1989.[19] Combat aircraft unit production costs appear to have been increasing by roughly 11 per cent per year in real terms since the early 1940s.[20]

Whether the product and process engineering costs of important private sector innovations have been propelled beyond the reach of any but very large corporations remains unclear. We know that projects in some fields have become extremely expensive. Average R&D and clinical testing costs for a new pharmaceutical chemical entity in the United States approach $100 million, and bringing a complete high definition television system into existence is said to require upwards of $700 million. But these may be exceptional cases, to be treated as such. And even for known big-ticket items, it is not completely clear that a large-scale organization is necessarily best. The most prominent counter example, which I cannot explain, was the 1989 decision of Cray Research Corporation to split itself into two separate and competing companies, one of which will stress evolutionary development of supercomputers using silicon semiconductors, with the other working on designs featuring riskier gallium arsenide chips.[21] Each development effort was expected to cost approximately $200 million.

We economists should be able to do more than cite extreme cases. A pertinent question in the evolving debate over firm size is whether there is a general tendency for civilian R&D to come in ever more expensive chunks. In an effort to shed light on the matter, I tapped the only relevant large-scale panel data set of which I know. Each year since 1963, a magazine, variously called *Industrial Research*, *Industrial Research & Development* and *Research & Development*, has sponsored a competition to name the most significant technological innovations of the past year. Since the late 1960s, the issues announcing the 100 winners have included information on mean and maximum R&D costs for the winning projects.

The annual maxima have varied in current dollar terms from $2 million (in 1970) to $960 million (in 1983, for a coal liquefaction process developed jointly by eight private and public corporations from four nations). When one analyzes the data systematically, one realizes why it is intrinsically difficult to make statements about averages and trends. The annual maximum R&D values (converted to constant dollars) exhibit a close fit ($r^2 = 0.974$) to a Pareto distribution with an alpha coefficient of -0.60, which means that asymptotically, the distribution has neither finite mean nor variance. When one samples from such a distribution, almost anything can happen, including the emergence of data patterns suggesting trends or cycles where none in fact exist.[22]

Throwing the cautionary message of this finding to the wind, I attempted to see whether there might be a systematic trend in the average constant dollar R&D cost of the annual competition-winning innovations. To avoid the overwhelming influence of outlying values, I eliminated from each annual average the cost of that year's most expensive innovation. The time trend regression using these purged data for the years 1969–86 had a positive but statistically insignificant slope, implying that R&D costs (averaging $1.35 million per innovation in 1982 dollars) rose at about 1.0 per cent per year. Although there are a host of reasons why this result might not tell a full or unbiased story, the evidence does not imply any general tendency for innovative R&D costs to become so high that only giant corporations need apply.

The wrong kind of D in R&D?

Another possibility, and one at the heart of the critique by MIT's Charles Ferguson, is that US corporations, and especially the small high technology venture firms, are doing the wrong kinds of R&D. They are very good, all acknowledge, at developing innovative *products*. Where they fail is in working out the details of production *process* technology so that costs are driven to minimum possible levels without sacrificing product quality.

This hypothesis is particularly striking in conjunction with evidence presented by Edwin Mansfield that on average, fifty leading Japanese corporations devoted 64 per cent of their R&D budgets to internal process development and improvement.[23] A matched sample of American firms was found to allocate only 32 per cent – half as much – to process R&D.[24] Putting two and two together, one must ask whether size enters into this difference. In particular, are large firms more process R&D-oriented than small innovators?

Although other information undoubtedly lies unexploited, I can produce two strands of evidence on this question. Keith Pavitt and associates have broken down their tally of nearly 4,000 UK manufacturing industry innovations *inter alia* into process innovation fractions by innovating firm-size class. The data are given in Table 2.1.[25] Consistent with the hypothesis, the smallest innovators are least process-oriented. The most process-oriented innovators are medium-sized companies,

Table 2.1   Proportion of process innovations by firm size.

Employees	Per cent process innovations
50 000 and more	33.9
10 000–49 999	31.2
1 000–9 999	37.2
200– 999	27.6
1– 199	26.9

with the largest firms in second place. The differences among size classes are not great, although the difference between the proportions for the largest and smallest sized cohorts is statistically significant.

My own research linking patented inventions to the R&D of companies covered by the 1974 Federal Trade Commission Line of Business survey yielded estimates *inter alia* of the fraction of R&D outlays that were process-oriented for 1,819 distinct lines of business with at least one linked patent. Pulling out a musty computer printout that included a correlation matrix, I discovered something of which I had previously been unconscious – that the process R&D fractions were positively correlated with the logarithm (to base 10) of the lines' 1974 sales, with $r = +0.24$. The correlation, though modest, is significantly different from zero at the 0.01 level.

A further analysis revealed that, unlike many such simple correlations between R&D performance variables and structural variables, this one did not disappear when inter-industry differences in technological opportunity were taken into account. In a multiple regression of the process R&D fraction (in ratio form) on the logarithm of 1974 sales, the four-seller concentration index CR4 (in percentage form) and intercept dummy variables differentiating "traditional" technology industries, organic chemicals, other chemicals, electrical equipment, electronic systems and components, metallurgical and non-manufacturing industries from the mechanical industries base case, the result was as follows:

$$PROCESS = 0.050 + 0.104\ LOGSALES - 0.0016\ CR4$$
$$(1.65)\quad (7.50)\qquad\qquad\quad (3.57)$$

$$+\ 0.31\ TRAD + 0.18\ ORGANIC$$
$$(12.83)\qquad\quad (6.27)$$

$$+\ 0.21\ OTHCHEM - 0.06\ ELECT$$
$$(7.44)\qquad\qquad (1.79)$$

$$-\ 0.00\ ELTRON + 0.34\ METAL$$
$$(0.02)\qquad\qquad (8.23)$$

$$+\ 0.30\ NONMFG \qquad R^2 = 0.202;\ N = 1819$$
$$(6.37)$$

with *t*-ratios for the coefficients given in parentheses. The *LOGSALES* coefficient reveals that each tenfold increase in line of business sales was accompanied on average by a 10 point increase in the percentage of R&D expenditures devoted to process innovation. Seller concentration, on the other hand, had a weak negative effect, with a 40 percentage point increase in the four-firm ratio reducing the process R&D percentage by 6.4 points.

Thus, there is apparently some substance to the proposition that large firms devote relatively more effort to process improvement than small firms. How important the difference is to the achievement of cost reductions, and how essential large scale is for innovators to take a serious interest in perfecting their production processes, remains unclear. My own view from qualitative observation is that differences in attitude may be much more important than differences in size. What is needed for good process R&D is a "hands on" approach to technological innovation, that is, with engineers spending a considerable fraction of their time on the production line talking with operators and worrying about how production methods can be improved. The training programs through which large Japanese corporations send their engineers seem ideally suited to inculcating such habits. In the United States, on the other hand, the graduates of top technical universities gravitate naturally toward industrial R&D laboratories set up to emulate academic ivory towers, and it is easy enough for people so sequestered to view "mixing it up" on the factory floor as dirty, noisy, demeaning activity, to be avoided whenever possible. I have consulted for a sizable US corporation that exerts strenuous efforts to combat such biases, leading to extraordinarily successful process innovation outcomes and superior productivity growth. If attitudes are the main cause of US firms' relative neglect of process work, merely encouraging the companies to become larger is unlikely to cure the problem, and by isolating the R&D specialists even more, it might aggravate matters. How one *does* alter attitudes is an extremely difficult question on which economists lack comparative advantage.

The changing parameters of patent litigation

Intellectual property rights are a third possible problem domain. Once relegated to the realm of esoterica, they have more recently become the subject of furious debate. There are questions as to whether, and to what extent, life forms should be patentable, how increasingly costly computer and semiconductor software should be protected, and much else. But here I wish to focus on a little-recognized question that could have important consequences for the small vs. large firm debate.

In 1983, the US Congress established in Washington, DC, a new Appellate Court for the Federal Circuit, charging it *inter alia* with responsibility for hearing all appeals on patent matters from the geographically decentralized district courts. The change from multiple to single court review of such technically specialized questions seemed a sensible one, and the expressed belief of the legislative drafters was that no substantive change in the law was intended. But the new court was "packed" with patent system advocates, and substance did change in a way that is at least partly reflected in the statistics. Someone once said that a patent is merely a license to litigate, and given this verity, the strength of patent protection depends upon the parties' ability to fight protracted patent suits and how often those who are attempting to enforce their hoped-for exclusive rights prevail in the courts. When enforcement efforts begin, the alleged infringers typically defend themselves by asserting that

(a) the patent is invalid; and
(b) even if it is valid, they had not infringed it under a correct interpretation of its claims.

Before the Federal Circuit Court's creation, these defenses were successful in more than two-thirds of the cases that reached the Appellate level. With a new pro-patent court calling the tune, the betting odds have been reversed. In recent years, approximately two-thirds of the patents challenged before the Federal Circuit Court are upheld as valid,[26] and from a sample of cases, it appears that about 53 per cent of the litigants prevail in proving that their patents were *both* valid and infringed. Thus, the balance of power in the United States has shifted toward patent holders and away from those attempting to circumvent patent restraints. The remaining question is, how will that power be distributed across the firm size spectrum? And in particular, will large firms be better able to exploit their superior litigating resources by suing small alleged infringers, with or without solid factual bases for doing so, and by virtue of the credible threat such litigation poses, impede the innovative efforts, or even threaten the very existence, of small high technology competitors?

To shed some light on this question, the records of 148 patent case decisions between 1983 and 1988 were analyzed by my research assistant, Gregory Smirin. "Small" firms were defined as those with sales of less than $25 million in the year of decision; "large" companies had sales above that threshold. "Winning" was defined to mean a determination that a patent holder had its patent upheld by the Federal Circuit Court as valid and/or infringed. Of the 63 cases involving "large" US patent holders, the patentees won in 34 cases, including 19 victories over large firms and 15 over small firms. Large patent holders lost in 16 cases to other large firms and in 10 cases to small firms, with three other cases not classifiable. Of the 71 cases involving "small" US patent holders, the small firms won in 42 cases, including 15 cases against large firms and a disproportionate 26 against other small firms. The small patent holders lost in 29 cases, including 13 against large firms and 15 to small firms, with two cases not classifiable. Overall, the analysis did not support an inference that small firms fared significantly less well against larger rivals in the set of cases that were litigated to the Appellate level. To be sure, only a small fraction of all alleged infringement cases are carried that far, and a small firm is unlikely to expend the resources required unless there is solid evidence in its favor. So it remains possible that the new balance of power will enable large patent holders to harass their smaller innovative brethren. It is also possible that small firms with a strong patent position can use the new system to fence out larger corporations able to make valuable process improvement contributions – although if that were so, the two parties should have a mutual interest in uniting their efforts through licensing or merger. Much remains to be learned on this set of issues, which could become increasingly important to the future balance among firms of diverse sizes.

2.5   Conclusion

Although I have spent an indecent fraction of my career studying the relationships between firm size and technological innovation, I conclude by confessing that there is much about which I remain uncertain. Certainly, the optimal mix of enterprise sizes is not cast in stone. It must evolve with the challenges posed by advancing science and technology. What we need to understand much better is how those challenges are changing. Is the cost of the "typical" R&D project, or the top 5 per cent of all projects, rising more rapidly than the ability of relatively small firms to secure financing, or is the opposite more nearly true? Do small firms perform certain functions poorly, but by their very innovativeness deprive large corporations of the opportunity to fill the void? Are the legal institutions of capitalism being modified in ways that encourage, or discourage, the most desirable pace of technical advance? These and other questions remain a fruitful domain for further research.

Notes

1. F. M. Scherer, "Firm size, market structure, opportunity, and the output of patented inventions", *American Economic Review*, vol. 55, December 1955, pp. 1121–2.

2. Quoted from the translation by William James Adams in "Firm size and research activity: France and the United States", *Quarterly Journal of Economics*, vol. 84, August 1970, p. 387 note.

3. See also Jean-Jacques Servan-Schreiber, *The American Challenge*, translated from *Le défi Américain*, 1967, New York: Atheneum, 1968, pp. xiii, 6, 43, 157–8, 159:

   > [W]e find an economic system that is in a state of collapse. It is our own. We see a foreign challenger breaking down the framework of our societies . . . . The greater wealth of American corporations allows them to conduct business in Europe faster and more flexibly than their European competitors . . . . What is most productive and decisive in the modern economy is the combination of the research factor with an industrial infrastructure, effective means of finance, and a large sales organization. The home office of a giant corporation coordinates all of these . . . . What should we do? In a nutshell: achieve a real economic union and build giant industrial units capable of carrying out a global economic strategy . . . . Only a deliberate policy of reinforcing our strong points – what demagogues condemn under the vague term of "monopolies" – will allow us to escape relative underdevelopment.

4. See Keith Pavitt, M. Robson and J. Townsend, "The size distribution of innovating firms in the UK: 1945–1983", *Journal of Industrial Economics*, vol. 35, March 1987, p. 304.

5. It is interesting that even though large firms out-innovated small firms in the United Kingdom during this early period, British scholars viewed the relative advantages of small vs. large enterprises in a more balanced way than the Fontainebleau conference participants. See, e.g., C. F. Carter and B. R. Williams, *Industry and Technical Progress*, London: Oxford University Press, 1957, pp. 184–7; John Jewkes, David Sawers and Richard Stillerman, *The Sources of Invention*, New York: St Martin's Press, 1959, p. 222; and Chris Freeman, *The Role of Small Firms in Innovation in the United Kingdom since 1945*, Research Report No. 6, Committee of Inquiry on Small Firms, London: HMSO, 1971.

6. See F. M. Scherer, *Innovation and Growth*, Cambridge: MIT Press, 1984, pp. 223–35; Zoltan J. Acs and David B. Audretsch, "Innovation, market structure, and firm size", *Review of Economics and Statistics*, vol. 59, November 1987, p. 568; and US National Science Board, *Science and Engineering Indicators: 1987*, pp. 36, 312. The National Science Board figures are biased because large firms have a higher probability of being included in the sample and, if included, the sales and R&D of both their unrelated and related lines are measured. But for the relatively few small firms with innovations, sales and R&D are no doubt highly specialized in the field from which the sampled innovation came.

7. William James Adams, "Firm size and research activity", *Quarterly Journal of Economics*, vol. 84, August 1970, pp. 394–406.

8. F. M. Scherer, "Technological change", p. 281, drawing upon a sample compiled by Gellman Research Associates. The French sample was

particularly small (only sixteen innovations) and hence subject to large firm size sampling errors.

9. See Christopher Freeman, *Technology Policy and Economic Performance*, London: Pinter, 1987, especially pp. 39–49.

10. David C. Mowery, "Industrial research and firm size, survival, and growth in American manufacturing, 1921–1946", *Journal of Economic History*, vol. 43, December 1963, pp. 953–79.

11. See "The values of Doriot", *The Economist*, June 20, 1987, p. 75; and the obituary of ARDC's founder Georges F. Doriot in the *New York Times*, June 4, 1987. Doriot taught the Harvard Business School course that profoundly influenced my own decision to study the economics of research and development.

12. Edward B. Roberts, "Entrepreneurship and technology", *Research Management*, vol. 11, July 1968, pp. 249–66.

13. As of 1966, Roberts traced thirty-nine new ventures to a single unnamed large electronics firm. The spin-offs' combined sales in 1966 were twice those of their former parent.

14. See "Venture adventures", *The Economist*, December 19, 1987, p. 67; and "Venture-capital drought", *The Economist*, 24 June 1989, p. 73.

15. US House of Representatives, Committee on Banking, Finance and Urban Affairs, Subcommittee on Economic Stabilization, Hearings, March 11, 1986.

16. See Albert N. Link and Laura L. Bauer, "An economic analysis of cooperative research", *Technovation*, vol. 6, 1987, pp. 247–60; and John T. Scott, "Diversification versus cooperation in R&D investment", *Managerial and Decision Economics*, vol. 9, 1988, pp. 173–86.

17. Remarks of Congressman Edwards of California, *Congressional Record*, February 21, 1989, p. H278.

18. Charles H. Ferguson, "From the people who brought you voodoo economics", *Harvard Business Review*, May–June 1988, pp. 55–62.

19. "Criticism mounts on Stealth cost", *New York Times*, 24 June 1989, p. 6. A substantial fraction of the cost was for software and tooling that will lessen the cost of production – *if* significant design changes are not found to be necessary.

20. William B. Burnett and F. M. Scherer, "The weapons industry", in Walter Adams, ed., *The Structure of American Industry*, eighth edition, New York: Macmillan, 1989.

21. "Cray splits operations into 2 rival entities", *New York Times*, 16 May 1989, p. D1; and "A computer star's new advance", *New York Times*, February 17, 1990, p. 35.

22. For an argument that observed long cycles in productivity growth could have similar purely stochastic roots, see the discussion by William D. Nordhaus in *Brookings Papers on Economic Activity*, "Microeconomics", 1989, pp. 323–4.

23. Edwin Mansfield, "Industrial R&D in Japan and the United States: A comparative study", *American Economic Review*, vol. 78, May 1988, p. 226.

24. My own estimate of the process R&D fraction for a larger 1974 sample was 24.6 per cent. See F. M. Scherer, "Using linked patent and R&D data to measure interindustry technology flows", in Zvi Griliches, ed., *R&D*,

Patents and Productivity, Chicago: University of Chicago Press, 1984, pp. 419–20.
25. Keith Pavitt, M. Robson and J. Townsend, "The size distribution of innovating firms in the UK: 1945–1983", *Journal of Industrial Economics*, vol. 35, March 1987, pp. 291–316.
26. Other estimates put the validation rate as high as 80 per cent. See "The battle raging over 'Intellectual Property'", *Business Week*, May 22, 1989, p. 79.

3

R&D, firm size and innovative activity

Zoltan J. Acs and David B. Audretsch

3.1   Introduction

The "Schumpeterian hypothesis" has mainly revolved around the issues of firm size and technological change. Because of assumed scale economies for research and development inputs in producing innovative output, it has been hypothesized that large firms have an inherent advantage in innovative activity.[1] Unfortunately, the lack of direct measures of innovative activity has forced researchers to infer indirectly whether such scale economies actually exist on the basis of the estimated relationship between firm size and R&D effort. As Scherer (1983a, pp. 234–5) reports, the empirical evidence suggests that "size is conducive to vigorous conduct of R&D." However, as Fisher and Temin (1973) and later Kohn and Scott (1982) demonstrated, the determination of an elasticity of R&D inputs with respect to firm size exceeding unity does not necessarily imply that scale economies exist for R&D in producing innovative output. This became clear in our 1987, 1988 and 1990 studies where we found that small firms can be at least as innovative as their larger counterparts in certain industries. Since it is widely known that the bulk of industrial R&D is undertaken by large firms, this was a particularly surprising result. Although the analyses were undertaken at the aggregated industry level, the results cast at least some doubt on the virtually untested but central proposition that scale economies exist for R&D in generating innovative activity.

The purpose of this chapter is to apply one of the first direct measures of innovative activity at the first level to determine whether scale economies do, in fact, exist for R&D inputs. While our previous work enabled us to identify the relationship between market structure and innovative activity at the industry level, here we are able to estimate a firm production function for innovative output based on R&D inputs. We combine firm records for innovative activity with similar R&D measures that have been used by Scherer (1965, 1977), Soete (1979) and others. Thus, rather than having to infer indirectly the relationship between innovative inputs and outputs at the firm level, we are able to estimate this relationship directly (see also Acs and Audretsch's forthcoming study).

In the second section we examine how the relationship between innovative inputs and firm size has been used to make inferences about the relationship between inputs and innovative output. In the third section a production model of innovative output based on R&D input is estimated. The relationship between firm size and innovative activity is examined in the fourth section. Finally, in the last section a summary and conclusion are provided. We find strong evidence that economies of scale do not play an important role in producing innovative output. This resolves the apparent paradox in the literature that large firms engage in most of the industrial R&D, and yet, small firms contribute at least a proportional share of patent and innovative activity.

3.2   Innovative inputs and outputs

Fisher and Temin (1973) argued that there are two relationships which have been at the core of the so-called Schumpeterian hypothesis regarding firm size and innovative activity. While most researchers have tested whether R&D inputs increase more than proportionately with firm size, Fisher and Temin demonstrated that the Schumpeterian hypothesis could not be substantiated unless it was established that the elasticity of innovative output with respect to firm size exceeds one. They pointed out that if scale economies in R&D do exist, a firm's size may grow faster than its R&D activities. Kohn and Scott (1982) later showed that if the elasticity of R&D input with respect to firm size is greater than unity, then the elasticity of R&D output with respect to firm size must also be greater than one.

The literature has revolved around these two elasticities mainly because it has been exceedingly difficult to accumulate evidence on an essential element of the Schumpeterian hypothesis: do scale economies exist for innovative output with respect to inputs? Because of the lack of direct measures of innovative output, most studies have tried to infer whether economies of scale in R&D exist by estimating the two relationships discussed above. The few exceptions have involved a handful of studies relating R&D inputs to a proxy measure of innovative output – typically the number of patented inventions. As Scherer

(1965) observed, patents may actually be a closer measure of inputs in the innovative process than of outputs. In addition, Scherer (1983b) has found that the propensity to patent varies considerably across industries,[2] while Mansfield (1984, p. 462) concludes that the propensity for those inventions that are indeed patented to result in innovations also varies across both industries and firm size:

> The value and cost of individual patents vary enormously within and across industries . . . . Many inventions are not patented. And in some industries, like electronics, there is considerable speculation that the patent system is being bypassed to a greater extent than in the past. Some types of technologies are more likely to be patented than others.

Still, most of what is known about the relationship between innovative inputs and outputs is based on using patent counts as a proxy measure for innovative output.

In one of the most important studies, Scherer (1965) used the *Fortune* annual survey of the 500 largest US industrial corporations. He related the 1955 employment of research and development employees for 352 firms to the number of patents registered in 1959. Scherer found no evidence that the number of patents increased more than proportionately with firm size. Using the US Federal Trade Commission's Line of Business data, Scherer (1984, Chapter 11) was again able to examine the relationship between R&D and patents, this time for 124 separate industries. He found evidence of increasing returns to scale for R&D in only 15.3 per cent of the industries, but decreasing returns in one-quarter of the industries. However, Scherer (1983b, pp. 115–16) warned that

> this evidence on patent output–R&D input relationships could have either of two rather different interpretations: that the largest firms in an industry generate fewer patentable inventions per dollar of R&D than their smaller counterparts, or that they choose to patent fewer inventions of a possibly proportionate or even disproportionate inventive output.

By comparing the direct measure of innovative activity we introduced at the industry level in our 1987 and 1988 papers with firm sales and company expenditures, we are able to estimate directly the relationships between R&D and innovative outputs, and between firm size and innovative activity. The measure of innovative activity is the number of innovations made by a firm which are recorded in the US Small Business innovation database. These data are described in detail in our 1988 study, but several points should be emphasized. The database was created by recording innovations listed in technology, engineering and trade journals in 1982. An innovation was defined as the commercial introduction of a new product, process or service. While our previous studies examined these innovations aggregated at the industry level, we use the individual firm records in this study.

Based on a sample from the entire database, it was determined that the innovations recorded in 1982 were the result of inventions made, on average, 4.2 years earlier. Following the example of Scherer's (1965) assignment of a four-year lag between the invention and the issuance of the resulting patent, we assume a two-year lag exists between the firm's R&D expenditures and the invention. Thus, the innovation measure corresponds roughly to 1975 firm R&D sales and data.

While Scherer used the data from the *Fortune* 500 in his 1965 study, we follow the example of his 1977 paper, along with a study by Soete (1979), in using the R&D and firm size measures from the *Business Week* sample of over 700 corporations for which R&D expenditures · play an important role. An important feature of this sample is that it includes more than 95 per cent of the total company R&D expenditures undertaken in the United States. Since the innovation data have never been previously used at the firm level, we introduce these data in Table 3.1, which compares the frequency of innovation for the thirty most innovative firms to firm size, measured by sales (in millions of dollars), company expenditures on R&D (in millions of dollars), and the R&D/sales ratio. Three important observations can be made from this table. First, the distribution of frequency of innovation is apparently skewed, with a few firms making numerous innovations, and most firms contributing just several innovations. In fact, of the 732 firms included in our sample, 306 contributed at least one innovation, while the remaining 426 did not produce a single innovation. Second, there is a concentration of the most innovative firms in several sectors of manufacturing – electrical equipment and electronics, instruments, computers and office equipment, and non-electrical machinery. Finally, there is obviously no precise correspondence between R&D expenditures and innovative activity, or between the firm R&D/sales ratio and the frequency of innovation. Nor is there a clear tendency for the number of innovations to increase along with firm size.

It is, of course, conceivable that the quality or significance of innovations is not constant across either firm size or with respect to R&D effort. However, using a sample of nearly 5,000 of the innovations, the quality of an innovation, based on four different significance levels, was compared between large and small firms (Acs and Audretsch, 1988). The distribution of innovation across the four different significance categories was found to be virtually identical for large and small firms.

Although the *Business Week* sample excludes the smallest enterprises, firms which can be considered to be relatively small are included

Table 3.1   Most innovative firms.

Firm	Number of innovations	Sales ($ millions)	R&D expenditure ($ millions)	R&D/ sales (%)
Hewlett Packard	55	981.0	89.6	9.1
Minnesota Mining & Mfg	40	3 127.0	143.4	4.6
General Electric	36	13 399.0	357.1	2.7
General Signal	29	548.0	21.2	3.9
National Semiconductor	27	235.0	20.7	8.8
Xerox	25	4 054.0	198.6	4.9
Texas Instruments	24	1 368.0	51.0	3.7
Pitney Bowes	22	461.0	10.5	2.3
RCA	21	4 790.0	113.6	2.4
IBM	21	14 437.0	946.0	6.6
Digital Equipment	21	534.0	48.5	9.1
Gould	20	773.0	23.1	3.0
Motorola	19	1 312.0	98.5	7.5
Wheelabrator Frye	18	332.0	2.0	0.6
United Technologies	18	3 878.0	323.7	8.3
Hoover	18	594.0	4.3	0.7
Honeywell	18	2 760.0	164.2	5.9
Rockwell International	17	4 943.0	31.0	0.6
Johnson & Johnson	17	2 225.0	97.9	4.4
Eastman Kodak	17	4 959.0	312.9	6.3
Data General	17	108.0	11.6	10.8
Exxon	16	44 865.0	187.0	0.4
Du Pont	16	7 222.0	335.7	4.6
Stanley Works	15	464.0	3.5	0.7
Sperry Rand	15	3 041.0	163.5	5.4
Pennwalt	15	714.0	15.7	2.2
North American Philips	14	1 410.0	22.5	1.6
Harris	14	479.0	21.1	4.4
General Motors	14	35 725.0	1 113.9	3.1
Becton, Dickinson	14	456.0	17.8	3.9

in the data. A number of firms have fewer than 500 employees, which is the standard used by the US Small Business Administration to distinguish small from large firms. Table 3.2, which reveals the distribution of firms according to size, indicates that slightly more than one-quarter of the sample is composed of firms with less than $100 million of sales. Thus, an advantage of using the *Business Week* sample over, say, the *Fortune* 500, is that we are able to include a much greater spectrum of firm sizes.

There is, however, an inherent bias that may affect the results when these *Business Week* R&D data are used. In compiling the survey, a materiality criterion is imposed, such that companies only need to report if their R&D is "material", which means greater than 1 per cent of sales. This results in only high R&D/sales companies reporting, which may tend to be larger firms. The effect of this censoring would then be to exclude low R&D performers among small firms from the sample,

Table 3.2   The distribution of firms according to size.

	Sales class (millions of dollars)						
	1–49	50–99	100–149	150–199	200–999	1 000–1 999	2000+
Number of firms	59	125	78	57	246	80	87
Percentage	(8.1)	(17.1)	(10.7)	(7.8)	(33.6)	(10.9)	(11.9)

Table 3.3   Means for sales, employment, R&D and innovations for innovative and non-innovative firms (standard deviations in parentheses).

	All firms	Innovative firms	Non-innovative firms
Sales (millions)	1 190.181	1 894.530	603.573
	(3 364.223)	(3 539.196)	(1 692.045)
Employment	21 660.514	36 028.621	9 694.240
	(54 379.544)	(84 303.474)	(15 482.413)
R&D (millions)	22.642	42.661	5.970
	(79.541)	(114.271)	(11.382)
R&D/sales (%)	2.116	2.573	1.735
	(2.297)	(2.210)	(2.301)
Innovations	2.423	5.333	0.000
	(5.244)	(6.713)	(0.000)

while including the high R&D performers, resulting in a twisting of the fitted regression line to the right and a tendency to indicate increasing returns where none exist. However, to the extent to which this censoring is consistent across all firm sizes and R&D levels, the results will not be biased.[3]

In fact, the traditional hypotheses concerning firm size and innovative activity are confirmed when comparing the size and R&D effort between the innovating and non-innovating firms included in our sample. As Table 3.3 shows, the innovating firms are more than three times larger, in terms of sales, or nearly four times larger, in terms of employment, than the non-innovating firms. Of course, since the innovating firms are substantially larger than their non-innovating counterparts, the R&D/sales ratio is only about 50 per cent greater for the innovating companies.[4] Thus, Table 3.3 implies that innovative firms are both larger and spend more on R&D than firms that do not innovate.

Previous analyses by Scherer (1977) and Soete (1979) have shown that R&D effort increases at least proportionately along with firm size for the firms included in the 1975 *Business Week* sample.[5] In order to make our data comparable with the Scherer and Soete results, we exclude the service sector and petroleum companies, as well as include only the sales of Western Electric rather than those of AT&T in Table 3.4, which shows the share of sales, employment, R&D and innovations

Table 3.4 Share of sales, employment, R&D and innovations according to firm size.[a]

Number of firms ranked by sales	Share of sales (%)	Share of employment (%)	Share of R&D (%)	Share of innovations (%)
First 4	15.33	13.90	23.25	5.28
8	22.06	21.78	31.18	6.63
12	26.46	26.01	36.19	8.39
16	30.19	29.74	39.17	9.34
20	33.65	32.72	43.82	11.91
30	40.80	39.27	51.71	17.32
40	46.41	44.40	56.95	23.27
50	51.09	48.96	60.22	26.39
100	67.49	64.68	73.21	36.60
all 661	100.00	100.00	100.00	100.00

Note: [a]Excludes firms in the petroleum and service sectors. Only the sales of Western Electric rather than those of AT&T are included.

according to firm size. As Soete found, the share of R&D accounted for by the largest firms slightly exceeds that of their share of sales, but their share of innovations is clearly considerably less. Thus, Table 3.4 suggests that the largest firms contribute at least a proportional share of R&D, but a less than proportional share of the innovative output.

3.3   R&D and innovation

As a first approximation to answering the question whether scale economies exist for R&D in producing innovative output, a simple production relationship of the type used by Bound *et al.* (1984) can be examined:

$$I = aRD^{\beta_1} \tag{3.1}$$

where I is the number of innovations made by a firm, a is a constant, and RD is the firm's expenditures on R&D. Estimating this function yields:

$$\ln I = 0.5268 + 0.2592\ln RD \qquad R^2 = 0.213, \ F = 81.815, \tag{3.2}$$
$$(6.551) \quad (9.045) \qquad\qquad n = 306$$

where the t-values are listed in the parentheses. As Baldwin and Scott (1987) summarize, numerous studies have repeatedly found that the technological environment, or what Scherer (1965) terms the technological opportunity class, has a substantial effect on the relationship between firm characteristics and innovative activity. One way of controlling for the technological environment is to include a measure of the appropriate three or two-digit 1975 R&D/sales ratio.[6] These relatively broad industry R&D/sales measures are appropriate to use for firm data, since most of the firms in the sample do not restrict output to a single four-digit industry. More importantly, R&D is generally not targeted within a specific four-digit industry but can typically produce an innovation over a wider spectrum of products. Inclusion of the sector-specific R&D/sales measure (IRDS) yields:

$$\ln I = 0.4389 + 0.2286 \ln RD + 0.1529 \ln IRDS \qquad (3.3)$$
$$\qquad\ (4.882) \quad (7.479) \qquad\quad (3.059)$$

$$R^2 = 0.221, \, n = 306, \, F = 40.618$$

The estimated elasticities of innovative output with respect to R&D input in Equations 3.2 and 3.3 is not so different from the elasticity between 0.32 and 0.38 for R&D and patents based on 2,582 firms estimated by Bound *et al.* (1984).[7] Of course, the omission of firms with no innovations in the logarithmic regressions may bias the results. Substituting an arbitrarily small value, 0.1, for zero in the innovation measure in Equations 3.2 and 3.3 yields elasticities of 0.50 and 0.48, respectively, for the entire sample of 732 firms. Both of these estimated elasticities are significantly less than unity.

While the hypothesis that H_0: $\beta_1 \geqslant 1$ cannot be accepted at the 95 per cent level of confidence for either Equation 3.2 or 3.3, it does not automatically follow that there are not increasing returns to scale for R&D over the range of firm size included in the sample. In both his 1965 and 1983 articles, Scherer warned aganist the validity of inferences from production relationships of the type represented by Equation 3.1. Problems arise from

(a) the observations with zero values which have to be omitted from the estimation,
(b) the violation of the homogeneity assumption for input–output relationships since firms can innovate without R&D expenditures, and
(c) the greater weight which is accorded to firms with smaller R&D expenditures in the logarithmic transformation.

As as result of these statistical problems, Scherer instead relies on the estimation of cubic relationships in his 1965 paper and quadratic relationships in his 1983b paper. Following this example, we estimate innovative output as a function of R&D with squared and cubic terms in Equation 3.1 of Table 3.5. One of the statistical concerns with estimating the function with ordinary least squares (OLS) is that there is a high frequency of non-innovative firms included in the sample. Thus, in Equation 3.2 the same cubic equation is estimated using the TOBIT method. The results are virtually identical to those obtained from the

Table 3.5 Regressions of firm innovations and R&D (*t*-statistics in parentheses).

	(1) OLS	(2) TOBIT	(3) TOBIT	(4) TOBIT	(5) OLS
RD	0.1047 (13.903)[a]	0.1904 (11.835)[a]	0.1755 (10.563)[a]	0.1158 (10.807)[a]	0.0807 (6.365)[a]
RD^2	-2.0439×10^{-4} (-7.536)[a]	-4.8420×10^{-4} (-7.142)[a]	-4.3977×10^{-4} (-6.396)[a]	-1.1692×10^{-4} (-7.940)[a]	-1.5959×10^{-4} (-4.723)[a]
RD^3	1.1136×10^{-7} (5.387)[a]	3.3010×10^{-7} (5.462)[a]	2.9559×10^{-7} (4.844)[a]	—	8.8020×10^{-8} (2.832)[a]
IRDS	—	—	0.3689 (3.747)[a]	4.1680 (4.162)[a]	0.3055 (2.927)[a]
Constant	0.7482 (4.335)[a]	-4.3809 (-11.711)[a]	-5.4386 (-10.546)[a]	-5.1513 (-9.928)[a]	1.8318 (3.184)[a]
R^2	0.297	—	—	—	0.236
F	102.732[a]	—	—	—	22.027[a]
Log–likelihood	—	-1251.216	-1177.179	-1188.980	—
Sample size	732	732	673	673	289

Note: [a]Statistically significant at the 95 per cent level of confidence, two-tailed test.

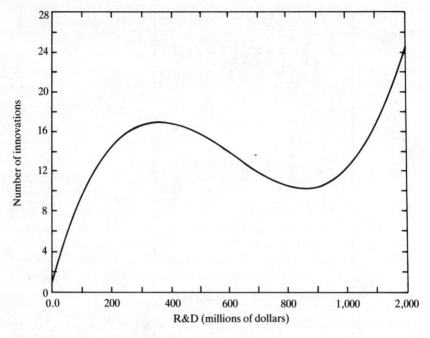

Figure 3.1 R&D and innovative output

OLS method. While the positive and statistically significant coefficient of the cubic term at first consideration might imply the existence of increasing returns from R&D, Figure 3.1 indicates that only two of the 732 firms are of a sufficient size to realize these scale economies – General Motors and IBM. Just as the logarithmic transformation places a greater weight on the smaller firms, the cubic function tends to exaggerate the importance of the largest firms. This is confirmed when the cubic term is omitted in Equation 3.4, leading to the conclusion that innovative output does not increase proportionately with increases in R&D. Thus, the appropriate interpretation of the cubic equations is that, with the possible exception of two firms, there is no evidence that economies of scale exist for R&D in producing innovations. This is similar to Scherer's (1965) findings.

This result is quite robust with respect to controlling for the technological opportunity environment and to adjustments in the firms included in the sample. For example, when the broad industry R&D/sales ratio is included in the model in Equation 3.3, the resulting curve looks quite similar to that in Figure 3.1. The second point of inflection changes only slightly from $705.3 million of R&D in Equation 3.2 to

$715.1 million in Equation 3.3. Similarly, the first inflection point changes only from $272.6 million of R&D to $276.8 million. An additional procedure for controlling for industry effects is to include thirteen dummy variables representing fourteen broad industry sectors. This produces results (using the TOBIT analysis) very similar to those in Equations 3.2 and 3.3 in Table 3.5:

$$I = -6.4940 + 13 \text{ Dummies} + 0.1618RD \qquad \text{(3.4)}$$
$$(-9.546) \qquad\qquad\qquad (11.794)$$

$$- 3.4025 \times 10^{-4}RD^2 + 1.9315 \times 10^{-7}RD^3$$
$$(-7.748) \qquad\qquad (5.31)$$

$$\text{Log-likelihood} = -1,229.34, \ n = 732$$

One concern might be that the presence of so many non-innovative firms might distort the estimated relationships. However, omitting the non-innovative firms from the sample in Equation 3.5 leaves the results almost unchanged. This is also true when the cubic term is dropped, producing a negative and statistically significant coefficient of the quadratic term.

Because of the consistent findings in the literature that the technological environment plays such an important role in the relationship between firm charactersitics and measures of technological change, the relationship between R&D input and innovative output was examined for each specific broad industry. As shown in Table 3.6, the results suggest that, in fact, the relationship between innovative inputs and outputs does vary substantially across industries. It should be noted that Table 3.6 includes only the specification (linear, quadratic or cubic), which proved to provide the best fit for each industry. All of the so-called high technology industries are found to exhibit evidence of decreasing returns to R&D, including an aggregation of the progressive industries, including firms in industries where R&D/sales ≥ 5 per cent, and the moderate industries, where 1 per cent < R&D/sales < 5 per cent. The resulting functions for the computer and chemical (with drugs) industries are shown in Figure 3.2. In fact, only one industry, petroleum, demonstrated evidence of increasing returns to R&D. An aggregation of the unprogressive industries, including firms where R&D/sales ≤ 1 per cent exhibited no evidence of increasing returns to R&D. These results generally confirm Scherer's 1983b findings of increasing returns to R&D in generating patented inventions in only 15 per cent of the industries. The existence of increasing returns to R&D in petroleum may reflect the tendency for innovative activity to require relatively large increments in knowledge in certain low technology industries. By contrast, in the high technology industries, relatively

Table 3.6 Industry-specific regressions for firm innovations and R&D (t-statistics in parentheses). (All equations are estimated using the TOBIT method.)

Industry	RD	RD^2	RD^3	Constant	Log–likelihood	n	Exceptions
Decreasing returns							
Chemicals (including drugs)	0.2218 (4.218)[b]	−0.0018 (−3.008)[b]	3.7260×10^{-6} (2.782)[b]	−1.3928 (−1.778)[a]	−179.702	84	Du Pont
Chemicals (excluding drugs)	0.1987 (3.492)[b]	−0.0016 (−2.648)[b]	3.4088×10^{-6} (2.628)[b]	−1.1536 (0.848)	−99.079	51	Du Pont
Machinery (excluding office)	0.6748 (1.360)	−0.0343 (−1.087)	4.0701×10^{-4} (1.023)	−3.0572 (−2.179)[b]	−74.685	51	Ingersoll-Rand
Computers & office equipment	0.1566 (3.462)[b]	-1.3990×10^{-4} (−2.999)[b]	—	−1.9877 (−0.939)	−69.590	31	—
Electrical equipment	0.6226 (4.430)[b]	−0.0047 (−3.338)[b]	9.1259×10^{-6} (3.102)[b]	−6.1123 (−4.105)[b]	−122.773	68	General Electric
Motor vehicles	0.0605 (2.383)[b]	-1.1714×10^{-4} (−1.895)[a]	6.7748×10^{-8} (1.781)[a]	−1.5893 (−1.616)	−49.719	31	General Motors
Instruments	0.6902 (5.448)[b]	−0.0020 (−5.019)[b]	—	−3.9245 (−2.487)	−101.930	46	—
Progressive industries[c]	0.2221 (7.280)[b]	-5.8407×10^{-4} (−4.931)[b]	3.9842×10^{-7} (4.031)[b]	−3.4221 (−3.927)[b]	−463.520	217	IBM

Table 3.6 continued

Industry	RD	RD²	RD³	Constant	Log–likelihood	n	Exceptions
Moderate industries[d]	0.0907 (6.607)[b]	-1.8221×10^{-4} (−4.234)[b]	1.0213×10^{-7} (3.353)[b]	−1.9801 (−5.095)[b]	−332.456	207	General Motors
Stone, clay & glass	1.7641 (2.066)[b]	−0.1844 (−1.818)[a]	0.0049 (1.671)[a]	−5.0125 (−3.076)[b]	−33.000	36	—
Increasing returns							
Petroleum	0.4100 (2.415)[b]	−0.0071 (−2.110)[b]	2.9360×10^{-5} (2.155)[b]	−5.1023 (−3.166)[b]	−52.487	39	Exxon
No conclusive evidence							
Unprogressive industries[c]	0.1369 (6.472)[b]	—	—	−4.2248 (−8.159)[b]	−365.599	262	—
Drugs	0.8331 (2.967)[b]	—	—	0.2622 (0.188)	−79.959	33	—
Foods & tobacco	0.2473 (2.855)[b]	—	—	−3.2991 (−3.417)[b]	−46.465	41	—

Notes: [a]Statistically significant at the 90 per cent level of confidence, two-tailed test.
[b]Statistically significant at the 95 per cent level of confidence, two-tailed test.
[c] Includes industries where R&D/sales $\geq$ 5.0%.
[d]Includes industries where 1% < R&D/sales < 5%.
[e] Includes industries where 1% $\geq$ R&D/sales.

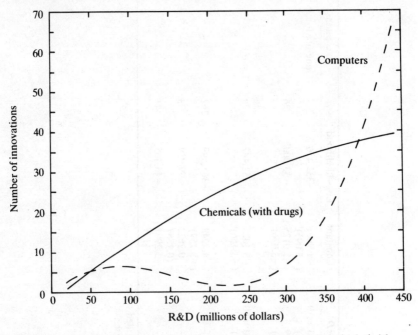

Figure 3.2 R&D and innovative output for computers and chemicals (with drugs)

small increments in the stock of knowledge will presumably yield an innovation.

3.4 Firm size and innovation

In some sense the relationship between firm size and innovative output can be considered to be a reduced form relationship consisting of two separate structural relationships – the relationship between firm size and innovative inputs, and that between inputs and innovative output. As Baldwin and Scott (1987) confirm in their thorough review of the literature, there has already been a plethora of studies examining the relationship between firm size and R&D effort, including a paper by Soete (1979) who compared the same R&D and firm size data we use in this paper. Although recent work by Bound *et al.* (1984) indicates that expenditures on R&D increase proportionately with firm sales, and Soete found that R&D increased more than proportionately with firm sales, virtually no one has found that this relationship is anything less than proportional.

However, just as there have only been a handful of studies examining the relationship between innovative outputs and inputs, the lack of meaningful data has not enabled researchers to estimate the relationship between firm size and innovative output. On the basis of the elasticity of R&D effort and firm size, which has been determined in empirical studies to be at least proportional, a similar pattern could be expected to emerge between innovative output and firm size. In light of our results in the previous section, however, the tendency of larger firms to be more R&D-intensive than their smaller counterparts may be offset by the lower productivity of that R&D. Thus, it remains to be empirically answered: to what extent does innovative activity increase or decrease along with firm size?

We again follow the example of Bound *et al.* (1984) in providing a first approximation of the relationship between firm size, measured by $ millions of sales (S), and innovative output by estimating a transformed logarithmic function:

$$\ln I = -0.0035 + 0.1750 \ln S \qquad R^2 = 0.079, n = 306, \quad \textbf{(3.5)}$$
$$(-0.016) \quad (5.113)$$
$$F = 26.15$$

The elasticity of innovative output with respect to firm size is statistically (at the 95 per cent level of confidence) less than unity, implying that innovative activity does not increase proportionately along with firm size. This result also emerges when an alternative measure of firm size, employment (E), is substituted for sales:

$$\ln I = -1.1389 + 0.2387 \ln E \qquad R^2 = 0.128, n = 306, \quad \textbf{(3.6)}$$
$$(-0.016) \quad (5.113)$$
$$F = 26.15$$

Including the measure of broad industry R&D/sales (IRDS) only slightly and insignificantly raises the elasticity:

$$\ln I = -0.4211 + 0.1957 \ln S + 0.2837 \ln IRDS \qquad \textbf{(3.7)}$$
$$(-1.785) \quad (5.683) \qquad (5.595)$$
$$R^2 = 0.163, n = 306, F = 27.950$$

Of course, the criticisms of the elasticity estimate based on simple logarithm transformations, which were discussed in the previous section, also apply to the relationship between firm size and innovative output. Therefore, as in the previous section, and as Scherer (1965) did in examining the effect of firm size on patents, a cubic function was

Table 3.7 Regressions of firm innovations and sales (t-statistics in parentheses).

	(1) OLS	(2) TOBIT	(3) TOBIT	(4) TOBIT	(5) OLS
Sales	0.0015 (8.161)[a]	0.0031 (8.542)[a]	0.0031 (8.513)[a]	0.0017 (7.195)[a]	0.0012 (3.765)[a]
Sales²	-8.7690×10^{-8} (−5.380)[a]	-1.8180×10^{-7} (−5.898)[a]	-1.8516×10^{-7} (−6.102)[a]	-3.434×10^{-8} (−4.834)[a]	-6.6317×10^{-8} (−2.505)[a]
Sales³	1.3779×10^{-12} (4.485)[a]	2.7876×10^{-7} (4.851)[a]	2.8818×10^{-12} (5.130)[a]	—	1.0212×10^{-12} (2.109)[a]
IRDS	—	—	0.7434 (7.008)[a]	0.7172 (6.628)[a]	—
Constant	1.0097 (4.878)[a]	−4.7448 (−10.496)[a]	−7.1549 (−11.505)[a]	−6.3557 (−10.527)[a]	3.6653 (7.676)[a]
R^2	0.120				0.071
F	32.987[a]				7.699[a]
Log-likelihood	—	−1303.611	−1208.960	−1222.24	—
Sample size	732	732	673	673	289

Note: [a]Statistically significant at the 95 per cent level of confidence, two-tailed test.

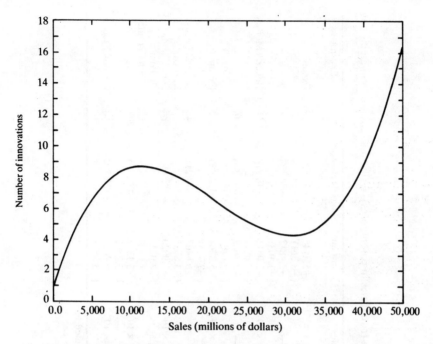

Figure 3.3 Firm size and innovative output

estimated. Equations 3.1 and 3.2 in Table 3.7 show that the coefficient of the cubic term is positive and statistically significant. This might seem to suggest that, for large firms, innovative output increases more than proportionately along with firm size. However, there are only two firms, Exxon and General Motors, whose sales exceeded the second inflection point of \$30,550 million in Figure 3.3. Thus, when the cubic term is omitted from the regression in Equation 3.4, the coefficient of the quadratic term is negative and statistically significant. This implies that the positive coefficient for the cubic term in Equations 3.1 and 3.2 is solely the result of the influence of two firms. The appropriate inference is, therefore, that innovative output increases less than proportionately along with firm size, with the exception of the two largest firms.

These results are quite robust with respect to controlling for the technological environment as well as for adjustments to the firms included in the sample. Including IRDS in Equation 3.3 only negligibly changes the estimated coefficient so that the second inflection point changes to \$31,113 million.[8] Similarly, including thirteen dummy variables in the regression to represent fourteen broad industrial sectors does not substantially alter the results (using TOBIT):

Table 3.8 Industry-specific regressions for firm innovation and size (*t*-statistics in parentheses). (All equations are estimated using the TOBIT method.)

Industry	Sales	Sales^2	Sales^3	Constant	Log–likelihood	n	Exceptions
Decreasing Returns							
Petroleum	0.0019 (2.887)[b]	-1.4205×10^{-7} (-2.520)[b]	2.4373×10^{-12} (2.514)[b]	-5.2145 (-3.361)[b]	-51.627	39	Exxon
Computers & office equipment	0.0083 (3.265)[b]	-4.6448×10^{-7} (-2.670)[b]	—	-2.4812 (-1.103)	-70.134	31	—
Motor vehicles	0.0020 (2.777)[b]	-1.2442×10^{-7} (-2.228)[b]	2.3009×10^{-12} (2.112)[b]	-2.0611 (-2.088)[b]	-48.218	31	General Motors
Instruments	0.0524 (5.723)[b]	-9.5606×10^{-6} (-5.386)[b]	—	-7.4682 (-3.940)[b]	-100.346	46	—
Unprogressive industries[c]	0.0026 (4.350)[b]	-2.1152×10^{-7} (-3.391)[b]	3.6805×10^{-12} (3.224)[b]	-5.3367 (-7.725)[b]	-372.023	262	Exxon
Food & tobacco	0.0057 (3.302)[b]	-9.0306×10^{-7} (-2.419)[b]	—	-6.6407 (-3.893)[b]	-40.211	41	General Foods Kraftco
Chemicals (including drugs)	0.0080 (3.918)[b]	-2.5097×10^{-6} (-2.813)[b]	2.4265×10^{-10} (2.541)[b]	-1.9020 (-2.023)[b]	-182.680	84	Dow Chemical Du Pont Union Carbide
Progressive industries[d]	0.0062 (5.490)[b]	-4.0316×10^{-7} (-2.718)[b]	6.8428×10^{-12} (1.708)[a]	-3.1222 (-3.268)[b]	-472.755	207	AT&T
Electrical equipment	0.0179 (4.278)[b]	-3.7124×10^{-6} (-3.264)[b]	1.9506×10^{-10} (3.063)[a]	-6.2017 (-3.882)[b]	-125.722	68	General Electric

Notes: [a] Statistically significant at the 90 per cent level of confidence, two-tailed test.
[b] Statistically significant at the 95 per cent level of confidence, two-tailed test.
[c] Includes industries where R&D/sales ≥ 1%
[d] Includes industries where 5% ≤ R&D/sales.

$$I = -6.4730 + 13 \text{ Dummies} + 0.0040S - 2.3148 \times 10^{-7}S^2$$
$$(10.887) \qquad\qquad (10.702) \quad (-7.748) \qquad (3.8)$$
$$+ 3.5779 \times 10^{-12}S^3 \qquad \text{Log-likelihood} = -1242.36,$$
$$(6.045) \qquad\qquad n = 732$$

When only innovative firms are included in the sample, Equation 3.5 indicates that the results are virtually identical to those including both innovative and non-innovative firms.[9] This also holds for a quadratic equation which was not included in the table. Thus, the results of the cubic and quadratic functions confirm the conclusion from the simple logarithm model: contrary to the Schumpeterian hypothesis, innovative output increases less than proportionately along with firm size.

Scherer (1965) found that the relationship between firm size and the number of patented inventions varies somewhat across broadly defined industries. However, Table 3.8 shows that decreasing returns to scale in producing innovative output emerge for every major industry.

It should be emphasized that although Soete (1979) used the same data to show that R&D expenditures increase at a more than proportional rate along with firm size, we find that innovative output increases less than proportionately with firm size. Thus, our results are more consistent with Scherer's 1965 finding that diminishing returns generally exist for the generation of patented inventions with respect to firm size.

3.5   Conclusion

Something of a paradox has emerged in the literature on technological change. While the Schumpeterian hypothesis, which has received at least some empirical validation, suggests that R&D inputs increase at more than a proportional rate along with firm size, Scherer (1965) has found that the generation of patented inventions increases less than proportionately with firm size. Using a new data source directly measuring innovative output, we are able to explain this apparent paradox. Although larger firms may be more R&D-intensive than their smaller counterparts, the productivity of R&D apparently falls along with firm size. There is no evidence that increasing returns to R&D expenditures in producing innovative output exist. Rather, the empirical results in this paper suggest, with few exceptions, diminishing returns to R&D are the rule. Thus, while larger firms are observed to undertake a greater effort towards R&D, each additional dollar of R&D is found to yield less in terms of innovative output.

One of the more confusing aspects in the literature concerning the relationship between technological change and firm size is that the conclusions are usually based on some truncated sample of firm sizes. It should be emphasized that the finding here of diminishing returns to R&D with respect to innovative output might emerge only in a sample of high R&D firms containing relatively few smaller enterprises. In any case, the results in this chapter make it clear that a firm need not reach gigantic dimensions in order to contribute to innovative output.

Notes

1.   For a review of this literature, see Baldwin and Scott (1987).
2.   According to Scherer (1983b, pp. 107–8),

> the quantity and quality of industrial patenting may depend upon chance, how readily a technology lends itself to patent protections, and business decision-makers' varying perceptions of how much advantage they will derive from patent rights. Not much of a systematic nature is known about these phenomena, which can be characterized as differences in the propensity to patent.

3.   As will be shown in the following sections, we find virtually no evidence for increasing returns, suggesting that the bias is not greatly affecting the results.
4.   The differences in means sales, employment, R&D and R&D/sales between the innovative and non-innovating firms are all significant at the 95 per cent level of confidence.
5.   It should be noted that the relationship between firm size and R&D expenditures within the 1975 *Business Week* sample is somewhat sensitive to adjustments in the data. When Scherer (1977) adjusted the data by eliminating firms in the publishing, restaurant and service sectors, and by using the sales of Western Electric rather than those of AT&T, a proportional relationship emerged. When Soete (1979) adjusted the data by eliminating the firms in the service sector along with the petroleum companies (because the sales data were greatly affected by the oil shock), the largest firms are found to spend a slightly greater than proportional amount on R&D.
6.   The three and two-digit R&D/sales ratios are from National Science Foundation, *National Patterns of Science and Technology Resources*, Washington, DC, 1985.
7.   Pakes and Griliches (1980) estimated a considerably greater elasticity between patents and R&D, 0.61, as did Hausman *et al.* (1984), 0.81. However, their samples were restricted to very large firms.
8.   The first inflection point falls from $11,617 million in Equation 3.1 to $11,452 million in Equation 3.2.
9.   In Equation 3.5 the first inflection point is $12,880 million and the second inflection point is $30,420 million.

References

Acs, Zoltan J. and David B. Audretsch, "Innovation, market structure and firm size", *Review of Economics and Statistics*, vol. 69, November 1987, pp. 567–75.

Acs, Zoltan J. and David B. Audretsch, "Innovation in large and small firms: An empirical analysis", *American Economic Review*, vol. 78, September 1988, pp. 678–90.

Acs, Zoltan J, and David B. Audretsch, *Innovation and Small Firms*, Cambridge, MA: MIT Press, 1990.

Audretsch, David B. and Zoltan J. Acs, "Innovation and size at the firm level", *Southern Economic Journal*, forthcoming (1991).

Baldwin, William L. and John T. Scott, *Market Structure and Technological Change*, London and New York: Harwood Academic Publishers, 1987.

Bound, John, Clint Cummins, Zvi Griliches, Bronwyn H. Hall and Adam Jaffe, "Who does R&D and who patents?" in Zvi Griliches, ed., *R&D, Patents and Productivity*, Chicago, IL: University of Chicago, 1984, pp. 21–54.

Fisher, Franklin M. and Peter Temin, "Returns to scale in research and development: What does the Schumpeterian hypothesis imply?", *Journal of Political Economy*, vol. 81, January/February 1973, pp. 56–70.

Hausman, Jerry, Bronwyn H. Hall and Zvi Griliches, "Econometric models for count data with an application to the patents–R&D relationship", *Econometrica*, vol. 52, July 1984, pp. 909–38.

Kohn, Meier and John T. Scott, "Scale economies in research and development: The Schumpeterian hypothesis", *Journal of Industrial Economics*, vol. 30, March 1982, pp. 239–49.

Mansfield, Edwin, "Comment on using linked patent and R&D data to measure interindustry technology flows", in Zvi Griliches, ed., *R&D, Patents and Productivity*, Chicago, IL: University of Chicago, 1984, pp. 462–4.

Pakes, Ariel and Zvi Griliches, "Patents and R&D at the firm level: A first report", *Economics Letters*, vol. 5, 1980, pp. 377–81.

Scherer, F.M., "Firm size, market structure, opportunity and the output of patented inventions", *American Economic Review*, vol. 55, December 1965, pp. 1097–125.

Scherer, F.M., *The Economic Effects of Compulsory Patent Licensing*, New York: New York University Graduate School of Business, 1977.

Scherer, F.M., "Concentration, R&D, and productivity change", *Southern Economic Journal*, vol. 50, July 1983a, pp. 221–5.

Scherer, F.M., "The propensity to patent", *International Journal of Industrial Organization*, vol. 1, March 1983b, pp. 107–28.

Scherer, F.M., *Innovation and Growth: Schumpeterian Perspectives*, Cambridge, MA: MIT Press, 1984.

Soete, Luc L.G., "Firm size and inventive activity: The evidence reconsidered", *European Economic Review*, vol. 12, 1979, pp. 319–40.

4

Firm size, university-based research and the returns to R&D

Albert N. Link and John Rees

4.1   Introduction

Scholars have become increasingly concerned about the role of small-sized firms in the innovation process. A number of important conclusions have come forth as a result of these inquiries.[1] First, small firms are more innovative (in terms of the number of product innovations) relative to their size than large firms. Second, product innovations coming from small firms appear to be more significant than those coming from large firms. Surprisingly, no studies to date have sought to explain, or even speculate, why small firms have this innovation-related advantage.[2]

This chapter compares university-based research relationships between small and large firms in an effort to identify one factor that might explain this noted difference in innovativeness. Our hypothesis is that innovation-based diseconomies of scale exist in large firms owing to the fact that bureaucratization in the innovation decision-making process inhibits not only inventiveness but also slows the pace at which new inventions move through the corporate system toward market. Small firms, who utilize university-based research relationships and are as a result more efficient in their internal R&D, partially avoid such problems.

This chapter is outlined as follows. In Section 4.2, the data that form the basis for our empirical investigations are described. In Section 4.3, we provide an overview of firms' involvement in university-based research relationships. The empirical analysis in Section 4.4 demonstrates that small firms are able to leverage their internal R&D activity through their research relationships with universities to a greater extent than large firms and, thus, enjoy a higher return on their research investments.

4.2   Description of the data

The sample of firms

In 1986/87 we assembled a data set related to firms' involvement with university-based research programs. Based on preliminary interviews with directors from both university and state research centers, several broad industry groups were identified to be the major "users" of such external research relationships. These industry groups included computing equipment, machine tools, and aircraft and components. From these broad industry groups, a population of 1,046 firms was identified from the 1986 DUNS file of the Dun and Bradstreet Corporation. After an initial mail survey to vice presidents of production/engineering, and follow-up telephone resurveys, complete information (defined below) was obtained on 209 firms.[3]

When surveyed, these firms were asked to classify themselves into one industry category based on their primary line of business. From their classification, these firms could be placed into five major SIC industry groups within the US manufacturing sector: metalworking machinery (SIC 354), office and computing machinery (SIC 357), electronic components and accessories (SIC 357), aircraft and parts (SIC 372), and engineering and scientific instruments (SIC 381). The distribution of firms across these five industry groups is shown in Table 4.1. Table 4.2 presents the distribution of these sample firms by size category. Along with the number of firms in each size category, the average number of employees per firm is also reported in that table.

Table 4.1 Distribution of sample firms by industry.

Industry	No. of firms
SIC 354	15
SIC 357	69
SIC 367	82
SIC 372	19
SIC 381	24
	209

Table 4.2   Distribution of sample firms by size category.

No. of employees	No. of firms	Av. no. of employees
<100	40	31
100 to 249	83	118
250 to 499	19	328
500 to 999	17	653
1 000 to 9 999	22	2 930
>10 000	28	76 556

Innovation-related characteristics of the firms in the sample

Although there are many ways to characterize the innovativeness of a firm, one dimension relates to self-financed R&D activity. The sample firms were classified as R&D-active or not, based on two separate criteria: R&D expenditures and R&D personnel. *A priori*, there was no reason to believe that these two indices would be perfectly correlated. For example, a firm that relies heavily on contracted research may not have an R&D budget proportional to its R&D staff. Likewise, especially in smaller firms, the R&D budget may be so insignificant both in absoulte and relative terms that the category "R&D personnel" is not meaningful. Or, the accounting system may not be refined sufficiently to separate R&D expenditures from other investments even when personnel are classified as related to R&D. Nevertheless, 93 per cent of the firms in this sample expended funds on R&D in 1986 and 88 per cent of all firms had at least one individual classified under the heading of R&D personnel. Table 4.3 shows the percentage of sample firms involved in R&D using each criterion. With the exception of the size category >10,000 employees, there is a marked similarity between the percentage of firms with an R&D budget and those with classified R&D personnel. In this largest category, 93 per cent of the firms reported an R&D budget but only 68 per cent reported having R&D personnel. Perhaps, and the data do not permit an investigation of this point, these largest firms rely most heavily on contracted research which is paid internally and conducted externally. For the entire sample of firms, the correlation coefficient between total R&D expenditures and total R&D personnel is 0.65 (significant at the 0.01 level or better).

Table 4.4 presents the percentage of sales devoted to R&D activity by size category for all firms in the sample. It appears that small firms devote a greater percentage of their sales to R&D than do large firms. While the percentage differences do not seem to be significant between the middle size categories, they are distinct between the category <100 employees and the category >10,000 employees. For all firms in the sample, the average per cent of sales allocated toward R&D is 10.6.

Table 4.3 Percentage of sample firms involved in R&D by size category.

No. of employees	$ (percentage)	Employees (percentage)
<100	88	83
100 to 249	93	90
250 to 499	95	95
500 to 999	100	100
1 000 to 9 999	100	95
>10 000	93	68

Table 4.4 Percentage of sales devoted to R&D by size category.

No. of employees	Percentage
<100	13.3
100 to 249	10.4
250 to 499	12.2
500 to 999	12.3
1 000 to 9 999	10.5
>10 000	5.0

Table 4.5 Percentage of total personnel involved in R&D by size category.

No. of employees	Percentage
<100	16.1
100 to 249	12.1
250 to 499	15.1
500 to 999	11.4
1 000 to 9 999	11.9
>10 000	7.9

A similar pattern across size categories for all firms in the sample is shown in Table 4.5. There, the percentage of total personnel involved in R&D decreases from 16.1 per cent in the category <100 employees to 7.9 per cent in the category >10,000 employees. The variation between the middle categories is again not striking.

4.3 Overview of firms' university-based research relationships

Sixty-nine per cent of the sample firms were involved with at least one university-based research program in 1986. As shown in Table 4.6, the degree of university involvement appears to increase with firm size. Whereas just over 50 per cent of the smallest firms (less than 250

Table 4.6 Involvement with university-based research programs by size category.

No. of employees	% of firms
<100	59
100 to 249	51
250 to 499	74
500 to 999	94
1 000 to 9 999	86
>10 000	100

Table 4.7 Involvement with university-based research programs by type of activity and by size category.

No. of employees	Consultants (%)	Contracts (%)	Research assts (%)
<100	41	15	47
100 to 249	44	22	38
250 to 499	67	44	72
500 to 999	76	29	59
1 000 to 9 999	77	54	64
>10 000	96	96	82

employees) were active in at least one research relationship with a university in 1986, about 90 per cent of the firms with more than 1,000 employees were so involved.

Information on three specific categories of involvement with a university-based research program was collected: faculty used as technical consultants (Consultants), contracted research projects (Contracts), and graduate students used as research assistants (Research assts). If a sample firm participated in at least one of these dimensions, then it was classified in Table 4.6 as involved in a university-based research program. The percentage of firms active in each of these three types of activities is shown in Table 4.7 by size category. In general, firms in the larger size categories make greater use of university faculty as technical consultants; however, in all three cases the percentage of firms in the size category >10,000 employees who are involved in any dimension is greater than for any of the other size categories.

The existing literature on industry–university research relationships suggests that these types of relationships are fostered by firms for two major reasons: it is a mechanism to reduce research costs and a method to identify potential productive employees. To investigate this issue further, each firm was asked to indicate which of the following were incentives (expected results from the relationship) to their participating in a university-based research relationship: "problem solving in production processes" (Pbl. sol.), "product development" (Prd. dev.), "use of university computing facilities" (Compt.), "use of other university facilities" (Facil.), and "gaining access to students as future employees" (Emplmt). The percentage of firms noting each of these as incentives is shown in Table 4.8 by size category.

With the exception of firms in the smallest two size categories, <100 employees and 100 to 249 employees, the potential for solving production process problems appears to be an important reason for firms to forge research relationships with universities. The importance of this potential as an incentive for such collaboration does not vary much by size category beyond firms with 250 or more employees. Over 60 per cent of the firms in the sample view product development as an important incentive for engaging in a research reltionship with a university. The use of university relationships as a vehicle to gain access to computing facilities appears to be primarily a small-firm (<500 employees) phenomenon. It may be the case that large firms have the in-house computer capabilities to conduct the requisite research operations. Access to other university facilities as an incentive for engaging in a university-based research relationship is important to some firms, but it does not seem to be systematically related to the size of these firms. In accordance with anecdotal information, gaining access to students as future employees is a significant incentive for firms of all sizes to pursue university-based research relationships.

The last column in Table 4.8 reports firms' responses to a question regarding the importance of "federal tax incentives as a motivation for engaging in collaborative research" with a university. While responses vary over size categories, only in the largest size category, >10,000 employees, did more than 50 per cent of the firms respond affirmatively.

Three response categories were used to determine firms' overall success with their university research relationships. The lion's share of the firms were satisfied with their collaborative research experience, as reported in Table 4.9.[4]

This overview of the primary data suggests several preliminary patterns of firm behavior. First, firms in all size categories were engaged in

Table 4.8 Incentives to engage in university research relationships by size category (in percentages).

No. of employees	Pbl. sol.	Prd. dev.	Compt.	Facil.	Emplmt	Tax
<100	3	61	5	23	55	24
100 to 249	19	69	18	19	65	16
250 to 499	57	71	29	57	79	40
500 to 999	37	63	6	19	69	29
1 000 to 9 999	42	47	5	16	84	11
>10 000	63	77	13	62	93	54

Table 4.9   Firms' overall success in university research relationships by size category.

No. of employees	'Very satisfied' (%)	'Somewhat satisfied' (%)	'Not satisfied' (%)
<100	29	64	7
100 to 249	38	62	0
250 to 499	71	29	0
500 to 999	44	56	0
1 000 to 9 999	46	54	0
>10 000	25	75	0

university research relationships to use, in general, faculty as technical consultants; and firms in the larger size categories do this to a greater degree than firms in the smaller size categories (see Table 4.7). This collaboration tends to be oriented primarily toward product development and secondarily toward problem solving in areas related to production (see Table 4.8). Second, in addition to research expertise, firms in all size categories viewed access to students as future employees as a significant incentive for engaging in a university-based research relationship (see Table 4.8).

4.4   The empirical analysis

This section presents the results of two empirical investigations into aspects of firms' participation in university-based research activity. In the first part, the trend noted in Table 4.6 that firms in the larger size categories were more active in university-based research relationships than firms in the smaller size categories is investigated statistically. The specific question considered is: is the probability of involvement with a university-based research program related to firm size? In the second part, the impact of external research relationships on the rate of return to firms' internal R&D is examined. Does the rate of return to R&D vary by firm size? Does the rate of return to R&D vary according to research participation with a university?

The propensity to engage in external research relationships

An inspection of the descriptive information in Table 4.6 suggests that the propensity to engage in a university-based research relationship is related to firm size. The percentage of firms active with a university-based research program increases with category size.

   To test formally for the influence of size on the propensity to engage in an external research relationship, a probit model was estimated. The independent variables in this model were firm size (*SIZE*) measured in terms of firm sales ($millions), industry concentration (*CR*)[5] and a binary variable equaling 1 if the firm was involved in basic research and 0 if it was not (*BASRES*).

The probit results, with asymptotic *t*-statistics in parentheses, are:

$$F^{-1}(P) = 0.114 + 0.016 \; SIZE \qquad\qquad (4.1)$$
$$\qquad\quad (0.31) \quad (2.94)$$

$$- \; 0.004 \; CR + 0.19 \; BASRES$$
$$(-0.44) \qquad\quad (0.63)$$

$$-2 \times (\text{log of the likelihood function}) = 185.92$$

where $P = F(\alpha + \beta_0 \; SIZE + \beta_1 \; CR + \beta_2 \; BASRES) = F(z)$ for F being the cumulative probability function. These regression results complement the pattern of activity shown in Table 4.6. The probability of participating in a university-based research program does increase with firm size.[6] The estimated coefficient on *SIZE* is significant at the 0.01 level or better. Industry concentration and involvement in a basic research program have no explanatory power in this specification.

Firm size, university-based research relationships and the returns to R&D

A framework frequently used by researchers in economics for estimating the returns to R&D reduces to the following regression model:[7]

$$TFPG = \alpha + \beta \; (RD/Q) + \epsilon \qquad\qquad (4.2)$$

where *TFPG* represents total factor productivity growth, (*RD/Q*) is the ratio of R&D spending to firm sales, and β is the estimated rate of return to R&D which could be interpreted as an index of R&D efficiency.

To estimate this model, data were needed for the calculation of total factor productivity over a defined period. Sufficient data for these calculations were not available for all firms in the sample. *TFPG* over the period 1982 to 1987 could be calculated for only 158 R&D-active firms of the 209 firms in the sample. The 51 firms deleted from the analysis were mostly small firms with <100 employees. Overall, the rate of return to the 158 firms in this subsample was 26.1 per cent. This result is reported in Table 4.10.

Several versions of the basic model were estimated. First, to test for differences between the rate of return to R&D in large versus small

Table 4.10   Estimated rates of return to R&D expenditures.

Category	Estimated rate of return (%)
Subsample of 158 firms	26.1
Large firms	26.0
Small firms	26.1
Firms involved in university research	34.5
Firms not involved in university research	13.2
Large firms	
Involved in university research	29.7
Not involved in university research	14.1
Small firms	
Involved in university research	44.0
Not involved in university research	13.9

firms, a second regressor was included in the above question. It equaled a binary variable interacted with the (RD/Q) term where the binary variable was given the value 1 for firms with less than 500 employees and 0 otherwise. The estimated least-squares coefficient on this term was not statistically different from zero, implying that there was no statistical difference between the returns in the two size groups.[8] This result is also reported in Table 4.10.

Second, a similar specification was estimated to account for possible differences in the rate of return to firms engaged in and not engaged in a university-based research relationship. For this, the binary variable equaled 1 if the firm was so engaged and 0 otherwise. As reported in Table 4.10, the estimated returns to R&D in firms involved in university-based research relationships are more than twice those of firms that are not – 34.5 per cent versus 13.2 per cent.[9]

Finally, segmenting both by size and by university involvement by including two regressors in the original specification (one with a size dummy and the other with a university-based research dummy), the returns to R&D in small, university-based research-active firms was found to be quite large. As reported in Table 4.10, the estimated rate of return to R&D in this group of firms was 44.0 per cent compared to

(a)   29.7 per cent in large university-based research-active firms,
(b)   14.1 per cent in large non-university-based research-active firms,
(c)   13.9 per cent in small non-university-based research-active firms.

Small firms appear to be able to transfer knowledge gained from their university research association most effectively, compared to large firms, to increase the returns to their internal R&D activities.

4.5 Conclusions

While the results presented in this chapter by no means explain fully why small firms have an innovation-related advantage over large firms, they do point out one interesting difference between an aspect of large and small firm research behavior. Although large firms are more active in university-based research *per se*, small firms appear to be able to utilize their university-based associations to leverage their internal R&D to a greater degree than large firms.

The analysis presented here did not take into account many of the other important factors associated with R&D efficiency, and so the results presented in Table 4.10 should be interpreted with caution. Still, the findings are noteworthy enough to encourage other investigators to investigate in more detail the ways in which firms internalize external technical information.

Notes

1. Much of this research is summarized in US Small Business Administration (1986) and in Link and Bozeman (1987).
2. Relatedly, Acs and Audretsch (1987a, b; 1988) show that the market environment most conducive for innovation is similar for both large and small firms. Also, they show that industry structure influences large firms' ability to innovate relative to small firms' ability to innovate, other things remaining equal.
3. Whenever possible reported survey information (e.g. sales data) was verified against published data (e.g. Form 10-K data) to insure response reliability. When explainable differences occurred (e.g. a survey respondent reporting sales in $millions rather than $thousands) the primary data were corrected.
4. There is not sufficient variation between the three response categories to conduct a more detailed investigation of inter-firm differences in success with university-based research.
5. $0 < CR < 100$. These data came from Weiss and Pascoe (1986).
6. A non-linear size variable was included in separate regressions, but the associated coefficient was not significant. As well, two other independent variables were considered. An index of foreign competition was included in a separate regression. This variable, based on data from the International Trade Commission and the Bureau of Industrial Economics, was constructed as the ratio of industry (four-digit) imports divided by the value of industry shipments plus imports less exports (Link and Bauer, 1989). It exhibited no statistical influence on the estimated probability. And, a vector of three-digit SIC industry dummies was included in the various versions of the model. As a group, these dummies were not significantly different from zero and thus were deleted. Similar results were obtained when *SIZE* was measured in terms of employees.

7. This model is fully explained in Griliches (1979). See Link (1987) for a review of the empirical literature. Using cross-sectional firm data from the US manufacturing sector, the estimated rate of return to internal R&D is in the neighbourhood of 20 per cent.
8. Link's (1980) analysis of the rate of return to R&D among firms in the chemicals industry found that the return increased with firm size to a modest threshold level, and then remained constant.
9. This finding may not be inconsistent with the findings of others that the returns to basic research are greater than for other categories of R&D spending. Generally, research conducted at universities is toward the basic end of the R&D spectrum; however, the underlying data (see Tables 4.7 and 4.8) do not allow us to separate clearly this form of research relationship. This finding also corresponds favorably with that of Link and Bauer (1989). They report, based on a sample of 92 manufacturing firms, the rate of return to R&D in firms involved in cooperative research programs with other firms is nearly three times that of firms not so involved – 37.7 per cent versus 12.9 per cent.

References

Acs, Zoltan J. and David B. Audretsch, "Innovation in large and small firms", *Economics Letters*, 1987a, vol. 23, pp. 109–12.

Acs, Zoltan J. and David B. Audretsch, "Innovation, market structure and firm size", *Review of Economics and Statistics*, vol. 69, 1987b, pp. 567–74.

Acs, Zoltan J. and David B. Audretsch, "Innovation in large and small firms: An empirical analysis", *American Economic Review*, vol. 78, 1988, pp. 678–90.

Griliches, Zvi, "Issues in assessing the contribution of research and development to productivity growth", *Bell Journal of Economics*, vol. 10, 1979, pp. 92–116.

Link, Albert N., "Firm size and efficient entrepreneurial activity: A reformulation of the Schumpeter hypothesis", *Journal of Political Economy*, vol. 88, 1980, pp. 771–82.

Link, Albert N., *Technological Change and Productivity Growth*, London: Harwood Academic Publishers, 1987.

Link, Albert N. and Laura L. Bauer, *Cooperative Research in US Manufacturing*, Lexington, MA: Lexington Books, 1989.

Link, Albert N. and Barry Bozeman, "Firm size and innovative activity: A further examination", final report to the Office of Advocacy, US Small Business Administration, 1987.

US Small Business Administration, *Innovation in Small Firms*, Washington, DC, 1986.

Weiss, Leonard W. and George A. Pascoe, *Adjusted Concentration Ratios in Manufacturing, 1972 and 1977*, Washington, DC: US Federal Trade Commission, 1986.

5
Technical performance and firm size: Analysis of patents and publications of US firms*
Alok K. Chakrabarti and Michael R. Halperin

This study focuses on the scientific output of firms of different sizes in different industries in the United States. Firm size has been of particular interest here as we have witnessed some recent major merger activities giving rise to consolidation and restructuring in many industries. According to Kamien and Schwartz (1975), beyond some magnitude, size does not appear especially conducive to either innovational effort or output. The Bolton Committee concluded that small firms contributed 10 per cent of all industrial innovations in the United Kingdom while accounting for only 5 per cent of R&D expenditure (Freeman, 1971). Very large firms, on the other hand, accounted for 54 per cent of R&D expenditure and accounted for 20 per cent of all innovations. Pavitt *et al.* (1987) have concluded that "companies with less than 1,000 employees commercialized a much greater share than is indicated by their share of R&D expenditure". Edwards and Gordon (1984) found that small firms produced 745 innovations per million employees while large firms generated only 313 innovations per million employees.

Although Edwards and Gordon found small firms to be more productive per employee in innovation, larger firms were more innovative in industries where four-firm concentration was 21–40%. Small firms were four times as innovative per employee as large firms in industries with concentration ratios of 41–60%. Acs and Audretsch

* Work in this area has been supported by the Science Indicators Unit at the Division of Science Resources Studies, National Science Foundation. We thank Dr Donald Buzzelli at the National Science Foundation for his continued support. An earlier version of this paper was presented at a research conference at the Science Center at Berlin, West Germany. Professor Wesley Cohen at Carnegie Mellon University has provided some helpful comments.

(1987) have supported the modified Schumpeterian hypothesis about imperfect market structure and benefits from innovation. Large firms seem to have innovative advantage in industries with a high level of concentration as well as capital and advertising intensity. Small firms derive innovative advantages in the early stages of the life cycle of an industry when total innovation and skilled labor play an important role. Acs and Audretsch (1988) found that the number of innovations increases with firm size at a decreasing rate. They also found unequivocally that innovation decreases with concentration.

A study of 130 largest US firms by Soete (1979a) has shown that firms tend to carry out proportionately more R&D with an increase in firm size, but at the same time tend to patent less. This finding supports an earlier finding by Scherer (1965).

We have focused on the technical performance of industrial firms, both large and small, as there has been increasing concern for the technological competitiveness of the United States (President's Commission, 1985). Several studies have pointed out that the innovative output of the United States has slowed down (Baily and Chakrabarti, 1988; Chakrabarti, 1988; Chakrabarti *et al.*, 1982; Gilsman and Horn, 1988). In this chapter we have investigated the technical performance of firms in different industries, both small and large, and the correlates of performance at a macro level. We have used both patents, and papers and publications as measures of technical output.

5.1   Industry samples

Small and medium-sized firms sample

Our sample is from the population of firms listed in COMPUSTAT database of public companies. Our criteria for sample selection were as follows:

(a)   annual sales in 1986 must be between $10 to $200 million;
(b)   annual R&D expenditure during the period from 1977 to 1986 must be at least $10 thousand; and
(c)   annual R&D expenses showed a growth of 10 per cent during this time period.

These criteria enabled us to obtain the sample of firms who were within a specific range of sizes and were involved in R&D in some serious and systematic way. We identified 248 companies by these criteria. We obtained data for these companies for the period 1977–86.

Large firms sample

The sample of large firms was developed from the list of public companies in the manufacturing sector having

(a)   minimum annual sales of $250 million
(b)   an annual R&D budget of $1 million
(c)   annual R&D budget at least 1 per cent of sales.

Using these criteria 225 companies were selected. We obtained data for these companies for the period 1975–83 (Halperin and Chakrabarti, 1987).

It may be mentioned that the studies reported by Soete (1979b) and Griliches (1984) used similar sources of data, i.e. R&D surveys by *Business Week*. We have stratified the sample by imposing the additional conditions as explained above.

Variables and sources of data

From COMPUSTAT we obtained the data on the following variables: sales, R&D expenditure, average sales during 1977–86, net income, and the main SIC number for the company.

The growth rates for sales, R&D and income are computed as the regression coefficient of the trend of the respective variables during the time period. The growth rate for income was not calculated for the companies which did not have consistent positive earning.

R&D expenditure data were obtained from *Business Week* for the group of companies in our large firm sample as that was the best source available at the time when we conducted the study in 1985–86.

Patent information

We used BRS/PATSEARCH, an on-line database to search for the patents issues to these firms from 1977 to date. This on-line database contains all utility patents, reissue patents and defensive patents issued by the US Office of Patents and Trademarks since 1975. To ensure complete coverage, we searched for the patents issued to the subsidiaries of these firms, if they had any. (See Halperin, 1986 for details about the technique of database search.)

Publications information

We obtained the information on papers and publications from SCISEARCH, a machine readable multidisciplinary index to the literature on science and technology prepared by the Institute of Scientific Information (ISI). All articles, reports, reports of meetings, letters, editorials, and correction notices from over 3,000 major scientific and technical journals are indexed in SCISEARCH. We obtained the number of papers and publications published by authors with an affiliation with the respective firms (or their subsidiaries) since 1972 to 1986 for the small firm sample. For the large firm sample the data on papers corresponded to the time period 1975–83.

Industry characteristics

We used the definition of Lawrence (1984) to categorize firms into five classes: high technology, capital intensive, labor intensive, resource intensive and service. High technology industries require a high proportion of R&D or employment of scientists and engineers intensively. Firms in capital intensive industries use standardized technologies and employ more capital than labor in production. Firms in labor intensive industries use labor intensive technologies. Firms in resource intensive industries use natural resources as their main input. We classified the sample firms into these categories by matching their four-digit SIC codes with those of Lawrence (1984, Appendix I).

5.2   Characteristics of the two samples

Table 5.1 provides the information on the firms in our two samples. The table indicates the difference between the two groups of samples. It is interesting to note that the large firms as a group have financially performed quite well in this period. Their growth rate in sales and income as well as R&D expenditure have been much better than the small and medium-sized firms. This points to the competitive stress in which the smaller firms have to operate. This table also points out that smaller firms have a better record of patents compared to their size, while the large firms tend to publish more.

Table 5.2 provides the data on the industry characteristics of the two samples of firms investigated in this chapter. Both samples had a large number of firms in the high technology industry. The sample for large

Table 5.1   Characteristics of the two samples.

Variables	Small firm sample (N = 248)	Large firm sample (N = 225)
Av. annual sales ($m.)	64.68	3 479.00
Av. annual R&D expenditure ($m.)	2.39	82.30
Annual income ($m.)	1.82	180.40
Annual R&D growth (%)	20.47	54.93
Annual sales growth (%)	10.09	47.78
Annual income growth (%)	11.17	39.45
Patent/co./year	2.12	50.46
Paper/co./year	0.55	43.29

Table 5.2   Industry characteristics of the two samples.

Industry	Small firm sample (N = 248)	Large firm sample (N = 225)
Hitech[a]	178	132
Capital intensive	28	28
Resource intensive	7	51
Labor intensive	26	Na
Others[b]	9	14

Notes: [a] Definition of these industries was according to Lawrence (1984). Hitech category included firms in chemical, electrical and electronics, machine tools and machinery, and instruments industries.
[b] "Others" included firms in the service sector such as oil services and health care.

firms had very few firms in the labor intensive industry. The sample for smaller firms, on the other hand, had only a few firms in the resource intensive industry.

5.3   Results

Correlates for research and development expenditures

R&D expenditure is closely related with the annual sales for both large and small firms. As shown in Table 5.3, this relationship is stronger in larger firms. Another important observation is that R&D budget is strongly correlated with income for only large firms.

Growth rate in R&D is linked with sales growth for all firms, the relationship is stronger in the case of smaller firms. Table 5.3 also indicates that smaller firms are likely to spend more money in R&D as their income grows. This is not so in the case of large firms.

Table 5.3   Correlates for R&D expenditures.

Variables	Small firm sample (N = 248)	Large firm sample (N = 225)
R&D budget		
Sales	0.43[b]	0.75[b]
Income	−0.11	0.83[b]
R&D growth		
Sales	0.78[b]	0.41[b]
Income	0.49[b]	−0.01

Notes: [a] Indicates significant at 0.05 level.
[b] Indicates significant at 0.01 level.

Table 5.4   Correlation between papers and patents.

	Small firms sample (N = 248)	Large firms sample (N = 225)
Correlation	0.32	0.79
Partial correlation controlling sales	0.29	0.39

Note: All correlations significant at the 0.01 level.

Inter-relationship between measures of R&D output

The two measures of R&D output, patents and papers are related to each other as shown in Table 5.4. The correlation coefficients among them are significant at the 0.01 level. We observe that the simple correlation between patent and papers for the large firms is higher than that for smaller firms. When we controlled for the size of the firms in our correlation analysis by computing the partial correlation, we find that the correlation is significantly lowered for the large firm sample, but not for the small firms. Moreover the partial correlation for these two samples are close to each other.

Correlates of patents

Table 5.5 provides the data on correlates for patents. Total number of patents and size are correlated as expected. We have two indicators of size, average sales and average R&D expenditure. For both the samples, the relationships are significant beyond the 0.01 level. However, for the sample of large firms, the relationship is stronger. For the large firms, patents and income are strongly correlated. We do not observe any statistically significant relations between income and patents for the small firms. Another important observation is that patents are not

Table 5.5 Correlates for patents.

	Correlation coefficients	
Variables	Small firm sample ($N = 248$)	Large firm sample ($N = 225$)
R&D expenditure	0.36[b]	0.64[b]
Sales	0.31[b]	0.56[b]
Income	0.10	0.60[b]
R&D growth	−0.11[a]	0.05
Sales growth	−0.06	0.06
Income growth	−0.18[a]	−0.01

Notes: [a]Significant at the 0.05 level.
[b]Significant at the 0.01 level.

Table 5.6 Correlates for papers and publications.

	Correlation coefficients	
Variables	Small firm sample ($N = 248$)	Large firm sample ($N = 225$)
R&D expenditure	0.44[a]	0.69[b]
Sales	0.14[a]	0.48[b]
Income	−0.03	0.67[b]
R&D growth	0.07	0.07
Sales growth	0.09	0.09
Income growth	−0.09	0.01

Notes: [a]Significant at the 0.05 level.
[b]Significant at the 0.01 level.

correlated with the growth related variables. If anything, we find a negative relationship between patent and R&D growth as well as income growth for the sample of small firms.

Correlates of publications

Table 5.6 provides the data on the correlates of papers and publications. Papers are correlated with R&D expenditure much more strongly than with sales for both large and smaller firms. These relationships are much stronger in the case of large firms. In the case of large firms, income and papers are strongly correlated. Interestingly, none of the growth variables are significantly related with papers.

Firm size and R&D productivity

Are small firms more productive than large firms? We have examined this question in our study. We have observed that in large firms both

Table 5.7 Comparison of concentration of papers, patents and sales (in percentages).

	Small firm sample			Large firm sample		
Size	Sales	Patents	Papers	Sales	Papers	Patents
First 4	8.31	3.24	2.32	23.70	7.20	7.90
First 8	13.96	5.71	3.21	35.20	25.70	19.60
First 12	17.02	8.75	3.45	43.20	29.70	24.90
First 16	20.84	10.06	9.77	48.70	30.70	28.20
First 20	24.52	10.89	10.66	53.30	35.60	34.70
First 30	33.02	22.63	22.50	62.40	48.20	50.00
First 50	47.44	39.56	33.74	74.00	73.80	61.30
First 75	61.61	50.15	43.56	83.00	79.20	72.80
First 100	72.69	63.32	64.87	89.60	88.50	81.00
First 150	88.05	81.12	79.63	97.30	98.20	95.70
All	100.00	100.00	100.00	100.00	100.00	100.00

Table 5.8 Firm size and R&D productivity for sample of smaller firms.

	Per $million R&D		
Size	Patent	Paper	Av. sales for the group
First 4	0.873	0.163	330.01
First 8	0.630	0.092	277.12
First 12	0.822	0.085	225.24
First 16	0.619	0.157	206.88
First 20	0.634	0.162	194.68
First 30	0.850	0.220	174.80
First 50	0.912	0.203	150.68
First 75	0.877	0.199	130.45
First 100	0.874	0.234	115.44
First 150	0.867	0.222	93.23
First 248	0.869	0.229	64.91

patents and publications increase less proportionately with sales for the largest 100 firms (Halperin and Chakrabarti, 1987). This finding is consistent with Scherer's (1965) findings. For the second sample of firms, however, we did not see such relationship. Table 5.7 provides the concentration data on the firms in our two samples.

To investigate the R&D productivity further, we computed the number of patents and papers per million of dollars of R&D for groups of companies of different sizes in our two samples. Table 5.8 provides the data on these variables for the small firm sample and Table 5.9 for the large firm sample.

We plotted the R&D productivity, both patent and papers, against the log of sales for the two samples. Figure 5.1 shows the plot for the large firm sample and Figure 5.2 for the small firm sample.

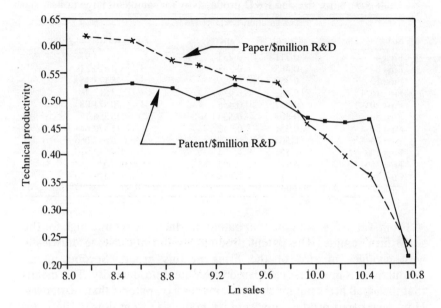

Figure 5.1   Technical productivity: large firms

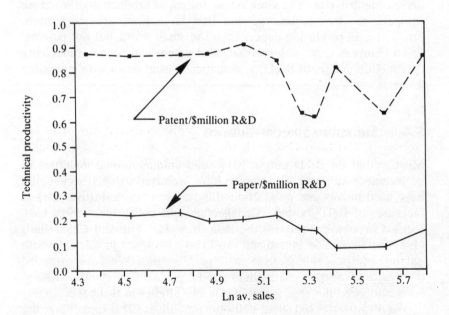

Figure 5.2   Technical productivity: smaller firms

Table 5.9 Firm size and R&D productivity for sample of large firms.

Size	Per $million R&D		Av. sales for the group
	Patent	Paper	
First 4	0.211	0.235	46 654.97
First 8	0.465	0.3633	4 429.31
First 12	0.460	0.396	28 127.15
First 16	0.462	0.434	23 812.38
First 20	0.467	0.456	20 824.08
First 30	0.503	0.534	16 256.77
First 50	0.531	0.542	11 570.64
First 75	0.504	0.565	8 657.066
First 100	0.524	0.574	7 017.915
First 150	0.532	0.609	5 079.14
First 225	0.526	0.618	3 479.610

From Table 5.8, we note that patent productivity is quite high for the small firm sample. The patent productivity did not change drastically with changes in size of the firms in this group. Second, paper productivity is much less compared with patent productivity. This means that the small firms put much more emphasis on patents than on papers. However, paper productivity went up as the size went down. These are represented in Figure 5.2.

In Table 5.9, we observe that patent productivity went up with decreasing firm size. The same is true for paper productivity. When we compare the two tables, we find that large firms are much more productive in producing papers than the small firms, but not patents. From Figure 5.1, we observe that both patent and paper productivity (per million dollars of R&D) systematically went down with increasing firm size.

5.4 Summary and conclusions

Most studies on R&D output have used unidimensional measures of performance indicator. Economists have preferred patents, sociologists have used papers and publications (and other communicative acts) as measures of R&D output. Management and policy researchers have focused on new products as the indicators. Halperin and Chakrabarti (1987) attempted to investigate the firms's behavior in terms of both patents and papers indicators in large US firms. We have extended that research to small and medium-sized firms in different industries and have compared the results from these two studies in this paper.

We have observed that determinants of R&D expenditure are different in firms of different sizes. For the large firms, R&D

expenditure depends on net income as well as its size measured in terms of annual sales. For smaller size firms, R&D expenditure is closely related with sales, rather than the net income. For the smaller-sized firms, the growth rate of R&D is closely related with growth rates of sales and net income. For large firms, growth rate in sales leads to a proportionate growth in R&D, but growth in income does not lead to proportionate growth in R&D. For larger firms, R&D is determined by its size as by available cash.

The two output measures, patents and papers, are correlated, but the strength of correlation is not a very great one for smaller firms. When we controlled for size and computed the partial correlation, we observe a much weaker relationship between papers and patents for the large firms. Our explanation is that there may be an inherent conflict between patents and papers. Patents and papers are probably substitute for each other. Scientists may not publish on an invention because that may be proved a hindrance to obtaining a subsequent patent on it (Halperin and Chakrabarti, 1987).

Patents and papers are correlated significantly with both R&D expenditure as well as annual sales. The strength of correlation is much greater for the larger firms than the smaller firms. This may be due to the fact that larger firms may encourage patenting and publishing more than the smaller firms.

The firm's growth is not linked with patent. If anything, we observe a negative relationship between patent and R&D growth, and patent and income growth in the case of smaller firms. Papers are not linked with growth variables for smaller firms. Rates of growth in both sales and income have significant correlation with R&D expenditure in small firms. For large firms, we observe a correlation between sales growth and R&D expenditure only.

Interpretation of the results of our study should take into account other variables and contextual conditions of the specific industries as well as firms. According to Rothwell and Zegveld (1982), smaller firms play a significant role in innovation in the earlier phase of a technology while larger firms benefit from process and other incremental innovations in the later stage of the technology's life cycle where economies scale become important in a mass market. Instead of providing a definitive conclusion about relative innovativeness or efficiency of smaller firms in the innovation process, we would echo the conclusions of Acs and Audretsch (1989) challenging the conventional wisdom of public policy favoring large firms.

The importance of the contextual conditions has been underscored by Levin and his associates in their survey of R&D executives in 130 industries in the United States. According to Levin *et al.* (1985), R&D

spending is encouraged in young industries where a strong science exists and where the government makes large contributions to technological knowledge.

A further analysis of the data from the small firms sample has shown that the relationship between R&D expenditure and patents and publications are different for different industries such as chemical, electrical, machinery and instruments (Chakrabarti, 1990a,b). The opportunities for patenting and the usefulness of patents as a competitive weapon vary from industry to industry. This may partly explain the differences in propensity to patent and publish. Focusing on the new products, it was found that the average number of new products generated per million dollars of R&D in the electrical industry was 2.078, while the same for the instrument industry was 1.158. This points to the need for a better understanding of the contextual conditions of the specific industries.

Are smaller firms more efficient than the large firms? The answer to this question is not clear. One of the major problems is that many small firms do not report R&D expenditures in the same way as the large firms. The problem is one of definition about what constitutes R&D. It appears to us that the more important question to examine is how the challenges of technological innovation are met by the small and large firms under different stages of the technology life cycle. As Chakrabarti (1989) has shown from a longitudinal study of three industries – chemicals, textiles and machine tools – productivity and innovation are related to each other. Different industries have dealt with the problems quite separately through the nature of R&D work undertaken by them, which in turn have been reflected in their technical output.

References

Acs, Zoltan J. and David B. Audretsch, "Innovation, market structure and firm size", *Review of Economics and Statistics*, vol. 69, no. 4, November 1987, pp. 567–74.

Acs, Zoltan J. and David B. Audretsch, "Innovation in large and small firms: An empirical analysis", *American Economic Review*, vol. 78, no. 4, September 1988, pp. 678–92.

Acs, Zoltan J. and David B. Audretsch, *Small Firms and Technology*, The Hague, Netherlands: Ministry of Economic Affairs, Directorate of Technology policy, 1989.

Baily, Martin Neil and Alok K. Chakrabarti, *Innovation and the Productivity Crisis*, Washington, DC: Brookings Institution, 1988, pp. 133.

Chakrabarti, Alok K., "Trends in innovation and productivity: The case of chemical and textile industries in the US", *R&D Management*, vol. 18, no. 2, April 1988, pp. 131–40.

Chakrabarti, Alok K., "Innovation and productivity: The case of chemicals, textiles and the machine tools industries in the US", *Research Policy*, vol. 19, no. 3, June 1990a, pp. 257–70.

Chakrabarti, Alok K., "Scientific output of small and medium size firms in high tech industries", *IEEE Transactions on Engineering Management*, 1990b.

Chakrabarti, Alok K., Stephen Feinman and William Fuentivilla, "The cross national patterns of industrial innovations", *Columbia Journal of World Business*, vol. 17, no. 3, Fall 1982, pp. 33–9.

Edwards, Keith L. and Theodore J. Gordon, *Characterization of Innovations Introduced in to US Market in 1982*, Washington, DC: US Small Business Administration Contract No. SBA-6050-A-82, 1984.

Freeman, Christopher, *The Role of Small Firms in Innovation in the United Kingdom since 1945*, Report to the Bolton Committee of Inquiry on Small Firms, Research Report No.6, London: HMSO, 1971.

Freeman, Christopher, *The Economics of Industrial Innovation*, London: Penguin Books, 1974.

Gilsman, Hans H. and Ernst Juergen Horn, "Comparative invention performance of major industrial countries: Patterns and explanations", *Management Science*, vol. 34, no. 10, October 1988, pp. 1169–87.

Griliches, Zvi (ed.), *R&D, Patents and Productivity*, Chicago, IL: University of Chicago Press, 1984.

Halperin, Michael R., *The Publications of US Industrial Scientists: A Company and Industry Analysis*, PhD dissertation, Philadelphia: Drexel University, 1986.

Halperin, Michael R. and Alok K. Chakrabarti, "Firm and industry characteristics influencing publications of scientists in large American companies", *R&D Management*, vol. 17, no. 3, July 1987, pp. 27–41.

Kamien, Morton I. and Nancy L. Schwartz, "Market structure and innovation: A survey", *Journal of Economic Literature*, vol. 13, March 1975, pp. 1–37.

Lawrence, Robert Z., *Can America Compete?*, Washington, DC: Brookings Institution, 1984.

Levin, Richard C., Wesley M. Cohen and David C. Mowery, "Appropriability, opportunity and market structure: New evidence on some Schumpeterian hypothesis", *American Economic Review*, vol. 75, No. 2, May 1985, pp. 20–4.

Pavitt, Keith L. R., Michael Robson and Joe Townsend, "The size distribution of innovating firms in the UK 1945–1983", *Journal of Industrial Economics*, vol. 35, no. 3, March 1987, pp. 297–316.

President's Commission on Industrial Competitiveness, *Global Competition: New Reality*, Washington, DC: Superintendent of Documents, 1985.

Rothwell, Roy and Walter Zegveld, *Innovation and the Small and Medium Sized Firm*, London: Francis Pinter, 1982.

Scherer, Frederick M., "Firm size, market structure, opportunity and the output of patented inventions", *American Economic Review*, vol. 55, no. 5, December 1965, pp. 1097–125.

Soete, Luc L. G., "The measurement of inventive activity and the firm size", Paper prepared for the Sixth EARIE Conference Paris, 9–12 September 1979a.

Soete, Luc L. G., "Firm size and inventive activity: The evidence reconsidered", *European Economic Review*, vol. 12, 1979b, pp. 319–40.

6

Formal and informal R&D and firm size: Survey results from the Netherlands

Alfred Kleinknecht, Tom P. Poot and
Jeroen O. N. Reijnen*

In the first part of this paper we compare the R&D data from the 1989 Stichting voor Economisch Onderzoek der Universiteit van Amsterdam (SEO) national survey on R&D and innovation in the Netherlands to the R&D data from the Dutch Central Statistical Office (CBS) as well as to a third independent source: the records on numbers of firms receiving subsidies on R&D. It will turn out that the SEO survey covers substantial amounts of small scale and rather informal R&D work which are lacking in the R&D survey by the CBS. This conclusion is backed by the independent evidence from the records on firms receiving R&D subsidies. It should be noted that the Dutch Central Statistical Office coordinates its data collection with the offices in the other Organization for Economic Co-operation and Development (OECD) member countries (in order to guarantee for international comparability). Therefore, our findings have implications for the reliability of data on R&D in small and medium-sized business in the entire OECD area.

In the second part, we explore the possible impact of the neglect of informal and small scale R&D in the official surveys, giving particular attention to the relationship of firm size and R&D, R&D concentration

* The authors are research fellows of SEO, Foundation for Economic Research of the University of Amsterdam, The Netherlands. We wish to thank Professor J.S. Cramer for his econometric advice, as well as Joost Hagens, Ineke Roest and Jorrit Verweij for their help in the administration and computer processing of the SEO national survey on R&D and innovation. The SEO survey has been financially supported by the Department of Technology Policy of the Dutch Ministry of Economic Affairs. The views expressed in this paper are solely those of the authors and are not necessarily identical with those of the Ministry of Economic Affairs or of SEO.

in large firms, and R&D and market structure. The third part covers our conclusions.

6.1 The underestimation of informal R&D in small firms

The SEO national survey on R&D and innovation in the Netherlands was held in 1989 (referring to data in 1988) among some 7,500 firms, having 10 or more employees. About half of these firms are in the manufacturing sector, the other half are service firms. The response rate was 58.1 per cent (4355 firms) and was quite even across sectors. However, smaller firms responded somewhat less than larger firms (see Kleinknecht *et al.*, 1990, for more details).

Already in an earlier (1984) innovation survey in the manufacturing industry of the Netherlands it appeared that there were many more small and medium-sized firms reporting R&D activities than was the case in the official survey by the CBS. Differences between the two surveys amounted to large orders of magnitude (Kleinknecht, 1987a). In the following, we shall document that the same holds for our 1989 survey. While the 1984 survey was confined to manufacturing industry, the 1989 survey also covers the service sector. As will be seen below, similar measurement differences hold for the service sector.

When analyzing the data from the 1984 survey, we only compared our data with those from the official survey. Now we have a third independent source, covering numbers of firms which receive R&D subsidies. In our recent survey, we also asked firms whether they had made use of R&D subsidies. This information from our questionnaire allows us to estimate percentages of enterprises per sector and size class which have done some R&D work but have not used R&D subsidies. Applying this latter information to the data set on numbers of firms which received R&D subsidies, we can estimate numbers of firms which do some R&D.

Table 6.1 compares the estimates from the three sources. It should be repeated that the response rate to our survey was only 58.1 per cent, and there are indications that innovative firms might have a higher propensity to respond to an innovation questionnaire than non-innovative firms.[1] We therefore extrapolated our R&D data in three different ways:

1. In a "minimum" estimate we calculate national totals under the assumption that firms which did not respond had no R&D.
2. In a "medium" estimate we assume that half of the non-responding firms had the same R&D behavior as the responding firms,

Table 6.1 Numbers of Dutch firms reporting R&D according to three sources, detailed by size classes and by manufacturing and services.

	Size classes (employees)						
	10–19	20–49	50–99	100–199	200–499	≥500	Totals
Estimate by Central Statistical Office (CBS)							
manufacturing	182	329	129	123	117	170	1050
services	312	245	67	37	50	55	766
totals	494	574	196	160	167	225	1816
Estimate based on R&D subsidy records							
manufacturing	565	941	818	546	264	246	3380
services	897	688	567	267	258	152	2829
totals	1462	1629	1385	813	522	398	6209
SEO ("medium" estimate)							
manufacturing	640	1087	697	434	262	101	3221
services	1401	1044	522	280	152	44	3443
totals	2041	2131	1219	714	414	145	6664

whereas the other half of the non-responding firms has no R&D.
3. In a "maximum" estimate we assume that all non-responding firms behave just like the responding firms, i.e. there is no bias problem.

In Table 6.1, we confine our documentation to the "medium" estimate which we judge to be the most realistic one. It is none the less remarkable to see from Appendix A that even the figures from the SEO "minimum" estimate still considerably exceed those from the official estimate by the CBS.

In interpreting Table 6.1, it should be noted that the numbers of enterprises according to the three different sources are not fully comparable: the CBS as well as the records on R&D subsidies count numbers of enterprises, whereas the SEO survey is based on "principal enterprise establishments".[2] An enterprise can have more than one principal establishment, which implies that the SEO survey counts slightly more units.

It is tempting to explain the differences between the three sources by arguing that the SEO survey counts more firms performing R&D because the respondents interpreted the definition of R&D more generously. Moreover, the records on numbers of firms applying for R&D could be inflated because firms redefined their R&D activities in order to increase their revenues from R&D subsidies.[3] We think that these arguments cannot explain the observed differences.[4]

It is certainly right that the description of R&D according to the Frascati Manual (as used in all three sources) leaves room for a "grey zone", which means that for certain activities it is difficult to decide whether they indeed fall under the Frascati definition of R&D. In the survey by the CBS as well as in the SEO survey, a shorthand Frascati definition of R&D on a separate sheet was sent along with the questionnaire. In order to reduce the chance of a "wide" interpretation of R&D, we included the following remark in the SEO questionnaire itself:

> On a separate sheet you find a (Frascati-Manual) definition of R&D. In questions 7–13 we deal with R&D in the mathematical, technical, medical, agricultural and natural sciences. Please note, in particular, that activities such as design and software do *not* belong to R&D.

This extra remark in the SEO survey questionnaire was intended as a precaution against an "inflationary" interpretation of the Frascati definition of R&D by the respondents. We have not been in a position to control among a substantial number of respondents whether they correctly interpreted the Frascati Manual definition (neither can such a control be carried out by the CBS). However, the institution which grants the subsidies for R&D usually applies such a control. They employ experts who control on each application form whether the applicant's description of R&D is consistent with the Frascati definition of R&D. Even if such a control cannot prevent some cheating by applicants, we think that the subsidy records are quite reliable. Therefore, we feel endorsed by the fact that the estimates from the R&D subsidy records come close to the SEO "medium" estimate.

What are then the reasons for the differences between the official estimate and the SEO estimate? In our view, the main reason is to be seen in the simplicity of our questionnaire. Unlike the questionnaire in the official survey, the SEO questionnaire is restricted to four simple questions (an English translation of these questions is given in Appendix B). It was deliberately decided to keep questions as simple as possible as we wished to achieve a high response rate. In order to keep things simple, we did not ask for R&D budgets but only for man years of R&D. There can be no doubt that man years are a crude measure as compared to money spent on R&D. The latter includes expenses for laboratory equipment and various other inputs in the R&D process. Moreover, budgets may capture differences in the quality of R&D workers as far as differences in qualifications and skills are reflected in wage sums.

On the other hand, our use of the most simple indicator had the advantage that small and medium-sized firms could respond more easily. Already from our earlier survey work we had the impression that there are many small and medium-sized firms which (occasionally) undertake small R&D projects. Such projects are often rather informal, i.e. there is no formal R&D department and perhaps not even a formal budget. Most of this R&D work may actually be "D" rather than "R". It is conceivable that such firms do not report their R&D in the official survey, either because they think it is not worth filling in a big questionnaire for such little amounts of R&D, and/or because their internal accounting system is not sufficiently detailed in order to give information on *budgets* (as opposed to time) spent on R&D. In other words, the complexity of the official R&D survey questionnaires as well as the fact that one asks for money spent on R&D may function as a filter against reporting small scale R&D activities.

The neglect of such small scale R&D in the official survey may explain a great deal of the differences in Table 6.1. As an illustration, Figures 6.1 through 6.4 give information about the percentages of firms which report a certain amount of R&D man years. Figure 6.1 covers the total sample of manufacturing firms, showing that 10 per cent of the firms report less than a quarter of a man year of R&D; 12 per cent report between one quarter and half a man year; some 14 per cent fall

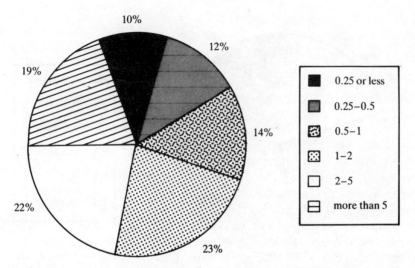

Figure 6.1 Percentages of firms reporting a certain amount of R&D man years in 1988 (only manufacturing firms, total national sample)

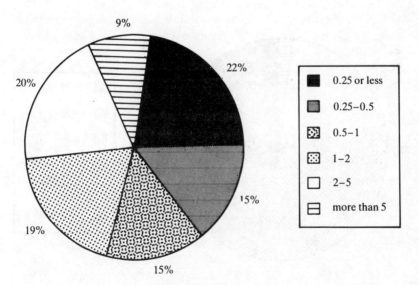

Figure 6.2 Percentages of firms reporting a certain amount of R&D man years in 1988 (only service firms, total national sample)

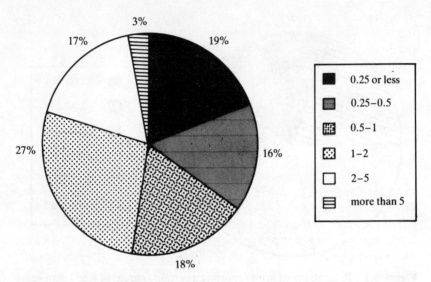

Figure 6.3 Percentages of firms reporting a certain amount of R&D (only manufacturing firms having 10–49 employees)

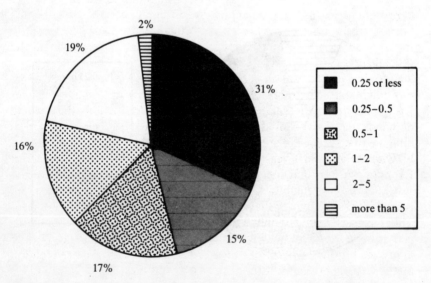

Figure 6.4 Percentages of firms reporting a certain amount of R&D (only service firms having 10–49 employees)

into the class between a half and a full man year, i.e. according to the SEO survey, more than one-third of Dutch manufacturing firms performing R&D have less than one man year of R&D. As can be seen from Figure 6.2, the share of small scale R&D performers is even higher in the service sector. More than half of Dutch service firms which perform some R&D report less than one man year. This is consistent with the evidence from Table 6.1 that differences between the official survey and the SEO survey are largest in the service sector. Looking at Figures 6.3 and 6.4, it comes as no surprise that firms having 10–49 employees have even higher shares of small R&D performers.

Unfortunately, it was not possible to extract comparable information from the official survey. However, we have the strong impression that the percentages of firms which report such small amounts of R&D are definitely lower in the official survey. Probably most of such small scale R&D is not captured in the official survey, which may explain much of the observed differences in terms of numbers of firms performing R&D in Table 6.1.

There are still other indications of an insufficient account of informal and small scale R&D in the official survey. Table 6.2 covers a ("medium") estimate of numbers of firms in the SEO survey which have a formal R&D department (first part). A comparison of the first row with the numbers of firms which have some R&D according to the official survey (second part) shows that both estimates hardly deviate from one another, at least when looking at totals (differences in size classes may in part be explained by differences in observed units: "firms" in the official survey versus "principal firm establishments" in the SEO survey). Therefore Table 6.2 confirms our impression that the official survey mainly captures *formal* R&D.

Looking again at the small amounts of R&D in Figure 6.1 through 6.4 one might raise the question of whether R&D in the smallest classes really falls under the definition of the Frascati Manual, as this manual demands that R&D should be a *continuous* activity. It is indeed hard to imagine that such small amounts of R&D can be performed continuously. One can also derive easily from the subsidy records of the Ministry of Economic Affairs that there are many smaller firms which receive R&D subsidies just in one subsidy period; then they disappear from the records and some periods later they come back with a new application. This suggests that, for many small firms, R&D is not a continuous activity but occurs rather occasionally. On the other hand, given that many small firms' applications for R&D subsidies pass the examination by the experts of the Ministry, we can say that their R&D indeed fulfils all the requirements of the Frascati Manual, except for the requirement of continuity. We therefore tend to argue that the R&D efforts of small

Table 6.2 Numbers of firms in the SEO survey which have a formal R&D department compared with numbers of firms which have R&D activities in the official survey.

	Size classes (employees)						
	10–19	20–49	50–99	100–199	200–499	>500[a]	Total
Firms having an R&D department (SEO "medium")							
manufacturing	95	327	291	234	162	72	1 181
services	117	155	94	88	58	14	526
totals	212	482	385	322	220	86	1 707
Firms having R&D (CBS)							
manufacturing	182	329	129	123	117	170	1 050
services	312	245	67	37	50	55	766
totals	494	574	196	160	167	225	1 816

Note: [a] In particular in this size class, figures can also differ due to different treatment of large concern enterprises (i.e. whether it was decided to have the questionnaire filled in by the holding company, by the division or at the company level).

Table 6.3 R&D man years by size class in the Netherlands. The SEO survey compared with the survey by the CBS.

	Size classes (employees)						
	10–19	20–49	50–99	100–199	200–499	>500	Total
SEO ("medium" estimate)							
manufacturing	640	2 415	2 164	2 479	4 005	23 510	35 213
services	1 081	1 646	1 541	1 951	1 496	1 979	9 694
totals	1 721	4 061	3 705	4 430	5 501	25 489	44 907
CBS estimate:							
manufacturing	260	480	450	1 020	25 180		27 390
services	440	450	240	120	1 800		3 050
totals	700	930	690	1 140	1 360	25 620	30 440

firms should not be disregarded, simply because they occur on a small scale.[5]

6.2   Exploring the impact of informal R&D

The impact on R&D concentration

The above-documented measurement differences do not only matter in terms of numbers of firms performing R&D work (Table 6.1), but also in R&D man years counted on a national scale. Table 6.3 compares the SEO "medium" estimate of man years of R&D (in 1988) with those from the official survey (in 1985 and 1986). There is only little difference in the size class of firms which have 500 and more employees. As far as the two estimates in this size class differ, we tend to have more confidence in the official estimate than in ours, as we had a lower response rate.

However, in the smaller-sized classes there are remarkable differences. Among firms which have 10–499 employees, we measure roughly four times as much, i.e. 19,418 man years as opposed to 4,820 man years in the survey by the CBS – a difference which also matters in terms of national totals (last column). However, as has already been mentioned (Note 4), the CBS data refer to 1985 and 1986, and there has been an increase of R&D work in the Netherlands in recent years (Kleinknecht *et al.*, 1990). As a consequence, the differences between the two estimates in Table 6.3, are likely to become smaller if, in future years, we will be in a position to compare the 1988 data of the CBS to the 1988 SEO data.

The two estimates in Table 6.3 also differ remarkably when looking at R&D concentration. According to the CBS survey (second part of Table 6.3), about 84 per cent of Dutch private R&D man years are done in firms which have 500 or more employees. According to the SEO estimate (first part), this figure is only 57 per cent. In other words, according to the SEO survey, the concentration of R&D in large firms is much less dramatic than is suggested by the CBS survey.

The impact on the relationship of market structure and R&D

Unfortunately, the most recent data on market structure in the Netherlands relate to 1981 and are confined to manufacturing industry. It seems to us therefore not appropriate to relate them to the R&D data from the recent SEO survey. Instead, we refer to an earlier investigation by Kleinknecht and Verspagen (1989) in which the data from the 1984 manufacturing survey were related to the 1981 market structure data. It turned out that a more adequate measurement of small scale R&D indeed makes a difference for the relationship between market structure and R&D.

In a first step, the data collectors at the CBS of the Netherlands were ready to estimate for us a regression of the R&D intensity (*RDI*) of forty-one branches on an indicator of market power (*ENTROPY*) in their database. This regression showed a highly significant linear relationship (see Equation 6.1). In interpreting Equation 6.1 it should be noted that *ENTROPY* is an *inverse* measure of market power (also known as the Theil coefficient), which implies that a negative coefficient of *ENTROPY* indicates a positive impact of market power on R&D.

$$\log RDI = 2.68 - 0.46\, ENTROPY \qquad t = 5.111 \qquad \textbf{(6.1)}$$
$$\text{adj } R^2 = 0.40 \qquad n = 41$$

The above equation from the official R&D database confirms Schumpeter's hypothesis that market power is conducive to innovation: the higher the degree of concentration, the higher is the R&D intensity. However, estimating the same regression on the same manufacturing branches with our R&D data set, market power turned out to be insignificant (see Equation 6.2).

$$\log RDI = 0.60 + 0.002\, ENTROPY \qquad t = 0.029 \qquad \textbf{(6.2)}$$
$$\text{adj } R^2 = 0.00 \qquad n = 41$$

In our view, the main reason for the insignificance of market power for R&D intensity in Equation 6.2 is related to our capturing of more small scale and informal R&D in small and medium-sized enterprises. To the extent that there are more small and medium-sized firms in markets with low levels of concentration, we might observe more R&D in such markets which reduces the slope of the regression line.

It can be added here that further data mining contributed only weak evidence in favor of a positive role of market power for R&D intensity. The most favorable case for market power emerged when splitting manufacturing into "modern" (high technological opportunity) and "traditional" (low technological opportunity) sectors. In the "traditional" sectors we found weak evidence (significant at a 90 per cent level) of a linear relationship between market power and R&D intensity. In the "modern" sectors there was somewhat stronger evidence (at 97.5 per cent level of significance) of an inverted U-curve (see Kleinknecht and Verspagen, 1989, for further details). It would still remain to be seen whether such evidence persists when controlling for other variables such as buyer (as opposed to seller) concentration, diversification, profitability, advertising intensity, etc. Recently, Scott (1984) and Levin *et al.* (1985) demonstrated that, even when using the conventional R&D sources, the importance of market power decreased drastically when controlling for such variables.

The impact on the relationship of firm size and R&D

Table 6.4 gives information about the importance of formal and informal R&D, i.e. about R&D taking place in a formal R&D department and R&D taking place without having an R&D department. It can be easily seen from the first column that, even according to our survey which captures much informal and small scale R&D, percentages of firms which undertake some R&D are strongly positively related to firm size. In manufacturing industry, only some 21 per cent of the firms (in services even only 12.4 per cent) in the smallest size class (10–19 employees) have some R&D activities against some 89 per cent (or 64.3 per cent in services) in the class of firms having 500 and more employees.

While the decision to undertake some R&D or to have a formal R&D department is positively related to firm size (Columns 1 and 2), the opposite seems to hold for the shares of informal R&D in total R&D (Column 4). In manufacturing, and a bit less so in services, the percentage of informal R&D is higher in smaller firms. In general, service industries are less R&D intensive than manufacturing sectors. The high shares of informal R&D in smaller firms as well as in service firms (as compared to manufacturing firms) may be taken as an indication that R&D in these two categories is rather a side activity than a "core" activity. This is also reflected in the "make-or-buy-decision": smaller firms as well as service firms have a higher share of their R&D performed by others via research contracts, whereas manufacturing firms in general, and larger firms in particular, have a higher share of in-house R&D.

Above we argued that the official R&D survey mainly captures formalized R&D. Table 6.2 shows that the number of enterprises which have a formal R&D department according to the SEO survey comes reasonably close to the estimate of numbers of firms reporting some R&D in the official survey by the CBS. In the following, we shall therefore use the R&D performed in an R&D department as an (admittedly rough) approximation of the R&D measured in the official R&D survey by the CBS. Of course, we could better have used the

Table 6.4   The importance of formal and informal R&D by size classes.

Size classes (employees)	Percentages of firms which have		Percentage of R&D which is done				n
	some R&D (1)	an R&D department (2)	in an R&D department (3)	without having an R&D department (4)	outside an existing R&D department (5)	externally via research contracts (6)	
Manufacturing							
10–19	21.2	3.5	27.7	50.4	2.1	19.8	174
20–49	34.6	9.5	39.8	34.4	8.7	17.1	442
50–99	62.6	22.2	44.0	30.9	7.5	17.6	677
100–199	76.9	38.8	53.7	15.4	11.2	19.6	430
200–499	86.3	47.1	63.3	13.3	13.0	10.5	281
≥500	89.4	64.5	77.9	1.7	10.3	10.2	153
Total manufacturing	42.4	15.4	66.6	11.2	10.1	12.0	2 157
Services							
10–19	12.4	1.1	14.2	47.0	4.4	34.4	566
20–49	14.0	3.0	44.4	33.6	6.5	15.6	566
50–99	26.8	3.2	29.9	44.2	4.4	21.4	339
100–199	46.5	6.6	28.6	19.1	2.1	50.2	280
200–499	48.9	10.9	38.7	34.5	12.5	14.1	253
≥500	64.3	22.6	54.1	17.0	9.0	19.8	145
Total services	17.9	2.9	37.6	28.6	6.8	27.0	2 149

R&D data from the official survey for this purpose. However, because of confidentiality reasons, it is impossible to get direct access to the latter.

In the following, we document several multiple regressions that were run in order to explore the impact of informal R&D on the relationship of R&D and firm size. All the equations were estimated in logarithmic form, applying weighted least squares (WLS) in order to deal with heteroscedasticity. Originally, these equations were estimated in order to assess the impact of various Dutch regions on the R&D intensity of enterprises. However, the regional dummies added to the equations proved to be insignificant and are left out here. In the equations below we present only those explanatory variables that have proven to have a significant influence on R&D intensity (*RDI*). We estimate the equations for manufacturing and service firms separately because of the considerable differences in R&D behavior (see Table 6.4) between these two groups. All regressions were run at the firm level, omitting firms which have zero R&D.

The independent variables are defined as follows:

SIZE   firm size (measured in numbers of employees)

IND   a dummy for independent firms (as opposed to firms which are part of a Dutch or a foreign concern)

EXP   a dummy for firms which export more than 50 per cent of their sales

HIGH   a dummy for manufacturing firms belonging to high technological opportunity branches (for a definition see Kleinknecht and Verspagen, 1989, p. 299)

SPEC   a dummy for service firms which belong to Sector 40 (government-owned gas, water and electricity supply firms) or Sector 975 (research labs)

COOP   a dummy for firms which have engaged in R&D cooperation with other firms or R&D institutions in the Netherlands or abroad

SALE   a dummy for firms which had a sales growth of 10 per cent or more in 1988 (as compared to 1987)

In the following equations, we estimate the influence of the variables on the R&D intensity of firms. In all equations, the R&D intensity of firms is defined as the sum of their R&D man years (intramural and/or externally) expressed as a percentage of all their employees. *T*-values are given in brackets. In Equations 6.3 and 6.4, the R&D intensity is based on *all* (formal and informal) R&D.

Equation 6.3 relates to manufacturing firms:

$$\log RDI = \begin{array}{cc} -3.029 & - & 0.277 \log SIZE + 0.477 \ EXP \\ (-22.699) & (-10.393) & (8.218) \end{array} \quad \textbf{(6.3)}$$

$$\begin{array}{cc} +0.494 \ HIGH + 0.359 \ COOP \\ (8.724) & (6.407) \end{array}$$

$$\text{adj } R^2 = 0.20 \qquad n = 1231$$

Equation 6.4 gives the same estimate for service firms:

$$\log RDI = \begin{array}{ccc} -1.976 & - & 0.513 \log SIZE + 0.199 \ IND \\ (-9.517) & (-14.300) & (-1.904) \end{array} \quad \textbf{(6.4)}$$

$$\begin{array}{ccc} +0.691 \ EXP + 0.385 \ COOP + 0.312 \ SALE \\ (4.378) & (3.637) & (2.912) \end{array}$$

$$\begin{array}{c} +0.520 \ SPEC \\ (2.643) \end{array}$$

$$\text{adj } R^2 = 0.35 \qquad n = 541$$

In both equations, the significant coefficients for exports confirm earlier findings by Hughes (1986) of a positive influence of exports on R&D intensity (implying a possibly simultaneous relationship). Also significant in both equations is R&D cooperation (*COOP*). In the manufacturing Equation 6.3, the significant dummy for *HIGH* confirms the importance of a distinction between high and low technological opportunity sectors. Within services, the dummy *SPEC* is significant as expected. One difference between the two equations relates to the *SALE* variable which proves insignificant in manufacturing, but significant in services. Moreover, as opposed to manufacturing firms, independent service firms have a lower R&D intensity than service firms which belong to a Dutch or foreign concern. For the purposes of this chapter, the most interesting finding in Equations 6.3 and 6.4 is the negative sign and the significance of the *SIZE* coefficient.

There seems to be some agreement in the literature that R&D increases *at least* proportionately with firm size. In other words, virtually nobody maintains that small and medium-sized firms are more R&D intensive than large ones; the majority of studies shows that large firms are more (or at least equally) R&D intensive than smaller ones (see e.g. the surveys by Kamien and Schwartz, 1982; Baldwin and Scott, 1987; Scherer, 1990). The behavior of the *SIZE* variable in Equations 6.3 and 6.4 clearly contradicts this common sense in the literature. In order to control whether our different outcomes have to do with the capturing of informal R&D in the SEO surey, we re-estimated the above equations, confining the R&D intensity to formal R&D only (*RDIF*); in other words, we only counted R&D taking place in a formal R&D department, trying to approximate the data from the official R&D survey.

R&D intensity (only formal R&D) in Dutch manufacturing firms (*RDIF*)

$$\log RDIF = \quad -2.868 \quad - \quad 0.291 \log SIZE + 0.239\, EXP \quad \textbf{(6.5)}$$
$$(-15.354) \quad (-8.469) \qquad (3.255)$$
$$+0.703\, HIGH + 0.200\, COOP$$
$$(8.881) \qquad (2.735)$$
$$\text{adj } R^2 = 0.25 \qquad n = 597$$

R&D intensity (only formal R&D) in Dutch service firms (*RDIF*)

$$\log RDIF = \quad -0.714 \quad - \quad 0.680 \log SIZE + 0.687\, COOP$$
$$(-2.163) \quad (-11.539) \qquad (3.737)$$
$$+1.308\, SPEC$$
$$(4.473)$$
$$\text{adj } R^2 = 0.57 \qquad n = 134$$

It is interesting to see that, notably in the service sector, the restriction to formal R&D caused a change of several coefficients. Some even became insignificant. However, to our surprise, the coefficients for the *SIZE* variable did not diminish. This was a reason to investigate more closely the R&D intensities of firms by size classes. Table 6.5 gives information about the R&D intensity of manufacturing and service firms by size classes according to the two definitions handled in the above equations (*RDI* and *RDIF*). This table shows a tremendously high R&D intensity of smaller firms, notably in the size class of 10–19 employees. The interpretation of these R&D intensities requires some words of caution. First, the R&D intensities are based on relatively small numbers of firms in our database, notably in the smallest size class. Second, these R&D intensities relate to the small minority of firms which report some R&D. As can be seen from Table 6.4 above, some 79 per cent of the manufacturing and some 88 per cent of the service firms in the smallest size class have *no* R&D and are therefore not included in the mean. Third, one should remind the very high standard deviation in the smallest size class which is also caused by the relatively small denominator. For example, a firm which has ten employees and one man year of R&D reaches an R&D intensity of 10 per cent; two man years would be 20 per cent, and so on. Fourth, as has already been mentioned above, response rates have been lowest in the smallest size class, and we cannot exclude that innovative firms responded more frequently than non-innovative ones. Fifth, Table 6.5 reports non-weighted means of the R&D intensities of firms. It should be noted that the Netherlands have a few large companies reporting high amounts of R&D. This implies that the (non-weighted) mean R&D intensities which are reported in the largest size class of Table 6.5 are

Table 6.5 The mean R&D intensity of firms by size class according to two definitions.

	Size classes					
	10–19	20–49	50–99	100–199	200–499	≥500
All (formal and informal) R&D						
manufacturing:						
mean[a] RDI[b] (%)	9.30	5.14	4.05	3.59	3.14	3.06
standard deviation	9.88	4.54	3.94	3.74	3.75	3.37
n (observation)	35	143	390	315	223	126
Services:						
mean[a] RDI[b] (%)	14.17	7.09	3.36	2.87	1.90	1.08
standard deviation	18.11	12.99	4.54	3.87	2.58	1.65
n (observation)	64	78	85	108	117	90
Only formal R&D						
manufacturing:						
mean[a] $RDIF$[c] (%)	17.20	6.40	4.68	3.55	3.40	2.86
standard deviation	14.65	4.07	3.65	3.53	3.87	3.08
n (observation)	6	43	163	168	128	90
Services:						
mean[a] $RDIF$[c] (%)	29.49	15.88	4.72	7.15	2.63	1.96
standard deviation	24.85	19.62	4.54	14.46	4.05	4.30
n (observation)	10	17	15	25	36	32

Notes: [a] The means are non-weighted arithmetic means of the R&D intensities of all firms which have some R&D (RDI) or which have a formal R&D department ($RDIF$).
[b] Definition of RDI: R&D man years (formal and/or informal) in 1988 as a percentage of all persons employed in an enterprise which has some R&D.
[c] Definition of $RDIF$: only R&D man years worked in formal R&D department as a percentage of all persons employed in firms which have an R&D department.

probably smaller than the R&D intensities which would result if we expressed the entire number of R&D man years in this size class as a percentage of total employment in the size class.

Although we omitted extreme outliers from our calculations, Table 6.5 suggests that our survey captured a number of small high tech firms, which have an extraordinarily high R&D intensity. The mean R&D intensity increases even further when restricting our observation to R&D that takes place in a formal R&D department which implies that we omit firms which often have only very small amounts of R&D. The very high R&D intensities in the smallest size classes obviously led us to find a negative coefficient of the *SIZE* variable in the above regressions. We therefore restricted our observations to firms which have 100 and more employees and then repeated the above estimates. The results are given in Equations 6.7 through 6.10.

Manufacturing firms ($\geq$100 employees), all R&D (*RDI*):

$$\log RDI = \begin{array}{cc} -4.124 & - & 0.091 \log SIZE + 0.378\,EXP \\ -(15.256) & (-1.890) & (4.820) \end{array} \qquad \textbf{(6.7)}$$
$$+0.589\,HIGH + 0.416\,COOP$$
$$(7.444) \qquad (5.394)$$
$$\text{adj } R^2 = 0.18 \qquad n = 638$$

Manufacturing firms ($\geq$100 employees), only formal R&D (*RDIF*):

$$\log RDIF = \begin{array}{cc} -4.791 & + & 0.020 \log SIZE + 0.319\,EXP \\ (-15.432) & (0.381) & (3.116) \end{array} \qquad \textbf{(6.8)}$$
$$+0.862\,HIGH + 0.158\,COOP$$
$$(8.088) \qquad (1.579)$$
$$\text{adj } R^2 = 0.21 \qquad n = 378$$

Service firms ($\geq$100 employees), all R&D (*RDI*):

$$\log RDI = \begin{array}{cc} -1.889 & - & 0.509 \log SIZE + 0.458\,EXP \\ (-4.350) & (-6.901) & (2.118) \end{array} \qquad \textbf{(6.9)}$$
$$+0.270\,COOP$$
$$(1.941)$$
$$\text{adj } R^2 = 0.15 \qquad n = 307$$

Service firms ($\geq$100 employees), only formal R&D (*RDIF*):

$$\log RDIF = \begin{array}{cc} -1.101 & - & 0.626 \log SIZE + 0.731\,COOP \\ (-1.499) & (-5.270) & (2.947) \end{array} \qquad \textbf{(6.10)}$$
$$+1.827\,SPEC$$
$$(4.396)$$
$$\text{adj } R^2 = 0.38 \qquad n = 90$$

It is remarkable to see from Equations 6.9 and 6.10 that the sign of the *SIZE* variable remains negative in services, implying that smaller service firms are, on average, *more* R&D intensive than larger ones. This holds independently of whether we include or exclude informal R&D in our measure of R&D intensity. In manufacturing, however, the distinction between formal and informal R&D matters. When including informal R&D (Equation 6.7) we obtain a negative coefficient for the *SIZE* variable in which case the *t*-value of −1.890 just fails to be significant at a 95 per cent level. When considering only R&D that takes place in a formal R&D department (*RDIF*) in Equation 6.8, we arrive at a result that is consistent with the common sense in the literature: the positive but insignificant coefficient of *SIZE* indicates that R&D intensity increases at least proportionately with firm size.

To conclude, the generally accepted hypothesis that R&D intensity increases *at least* proportionately with firm size can only be confirmed when confining our observations to manufacturing firms which have 100 or more employees, and when excluding informal R&D from our measure of R&D intensity. When including informal R&D and/or when considering smaller firms in our regressions, we find evidence that R&D intensity increases *less* than proportionately with firm size, i.e. smaller firms appear to be more R&D intensive than their larger counterparts, in particular in services.

6.3 Summary and conclusions

The simplified questions on R&D in the SEO national survey in the Netherlands lead us to find many firms reporting small scale R&D work which takes place in an informal way, i.e. without having a formal R&D department. The second part of this chapter shows that an attempt at capturing such informal R&D does indeed matter for subsequent analyses. It turns out that not only national totals differ; R&D concentration in large firms is also less than is suggested by the figures from the official survey. Moreover, the role of market structure for R&D is less important than when judging from the official CBS survey.

Perhaps the most important message comes from the behavior of the *SIZE* variable in Equations 6.3–6.10. In most of these equations we find evidence that smaller firms have a higher R&D intensity than larger ones. The tendency that R&D intensity increases *less* than proportionately with firm size is particularly strong in services. In several equations we tried to imitate the official survey by counting only the R&D work done within a formal R&D department. Of course, this is only a rough approximation, as the official survey also covers some firms which report R&D without having a formal R&D department. To our surprise, in several cases the *SIZE* coefficient does not change essentially when omitting informal R&D from our measure of R&D intensity. However, in the case of manufacturing enterprises which have 100 or more employees, the omission of informal R&D leads us to reproduce the finding which is generally accepted in the literature: R&D increases *at least* proportionately with firm size. However, when including *all* R&D, i.e. also the R&D work done outside formal R&D departments, we again obtain significantly negative coefficients. This is challenging the conventional wisdom.

It should be repeated that our regressions were run on firms which have *some* R&D, and that percentages of firms which have some R&D are positively related to size. Table 6.4 shows that, for example, in the size class of 10–19 employees, only 12.4 per cent of the service firms (and 21.1 per cent of the manufacturing firms) have some R&D. This implies that (even according to the SEO survey) only a minority of small firms reports any R&D activity. However, this minority tends, on average, to be more R&D intensive than larger firms. The only exception to this relates to firms which have 100 or more employees, in which case we find that R&D intensity is roughly proportionate to size when confining our observation to formal R&D. However, when including informal R&D, we again find a slightly negative impact of size on R&D intensity.

These findings also have implications for research which relates the R&D input of firms to some measure of "output" such as patents or numbers of innovations. The use of the official R&D data would lead one to strongly overestimate the efficiency by which small firms (as opposed to large firms) use their R&D resources. In general, our results are not of purely academic relevance. They also have some obvious policy implications, considering e.g. the attitude of governments towards mergers and take-overs, or towards subsidizing innovation in small or large enterprises.

The above should be a reason for R&D data collectors at the various institutions in the OECD area to reflect about a more adequate capturing of small scale and informal R&D in their surveys. Revisions of survey procedures need not necessarily be linked to the ongoing discussions on the revision of the Frascati Manual, since the measurement differences emerged in spite of handling an identical (Frascati) definition of R&D in all three sources considered above. Moreover, in our view, one can leave the R&D surveys among larger firms (having 500 and more workers) as they are. However, smaller firms should be approached with a radically simplified questionnaire. The copy from our questionnaire in Appendix B might serve as an example.

Without doubt, such a new approach would help to capture

considerable amounts of hitherto unobserved R&D which would cause a break in R&D time series. In an initial year, one might split the sample of small firms randomly into two parts, sending the new and simplified questionnaire to one half and the old questionnaire to the other half. This would help to produce valuable information for repair work on R&D time series later on.

There is yet another reason why a change in survey methods can be recommended. In recent years, a number of OECD countries have created facilities for subsidizing R&D, notably in small firms. As R&D is subsidized, firms may become increasingly aware that what they are doing is indeed R&D. A growing number of firms may therefore begin to report their R&D in the official surveys. As a consequence, the official surveys might report inflated growth rates of R&D. This is likely to have two advantages. First, the inflated growth rates in the official surveys will contribute to make the gap between the SEO survey and the official surveys gradually smaller. Second, government officials will be delighted to see in such growth rates a clear indication that their policy of subsidizing R&D indeed works.

Appendix A

Numbers of Dutch firms reporting R&D according to the SEO "minimum" and "maximum" estimate (to be compared to the SEO "medium" estimate in Table 6.1).

	Size classes (employees)						
	10–19	20–49	50–99	100–199	200–449	≥500	Totals
SEO "minimum" estimate							
manufacturing	482	807	509	326	204	86	2 414
services	1 039	741	392	210	122	34	2 538
totals	1 521	1 548	901	536	326	120	4 952
SEO "maximum" estimate							
manufacturing	797	1 366	886	542	320	117	4 028
services	1 763	1 346	652	349	180	55	4 345
totals	2 560	2 712	1 538	891	500	172	8 373

Appendix B

English translation of questions relating to R&D in the SEO national survey on R&D and innovation.

II Research and development (R&D)
On a separate sheet you will find a (Frascati Manual) definition of R&D. In Questions 7–13 we deal with R&D in the mathematical, technical, medical, agricultural and natural sciences.
Please note, in particular, that activities such as design and software do **not** belong to R&D.

NB   It may well happen that you have to answer a number of questions with "no" or "not relevant". For our research it is none the less important that you do answer these questions.

7   Has your enterprise a formal R&D department?
☐ **yes** (continue with question 8)
☐ **no** (continue with question 9)

8   If your enterprise has a formal R&D department: How many man years of R&D have in 1988 been spent on R&D within that department?
Man years of R&D in 1988: . . . .

9   R&D activities may be undertaken within your firm by departments other than an R&D department. For example: the sales department may develop a new product, or the production department may realize a process innovation.
Have R&D activities been undertaken during 1988 within your enterprise in departments other than an R&D department?
☐ **no** (continue with question 10)
☐ **yes.** If so, could you then give an estimate (if necessary a **rough** estimate) of the number of man years spent on such R&D in 1988?
Number of man years of R&D in 1988: . . . .

10   **External R&D:** Did your enterprise besides (or instead of) your own R&D also use external (domestic or foreign) R&D facilities? Examples are: TNO, universities, private R&D companies, other enterprises (please do **not** include contracts with parent, daughter or sister companies).
☐ **no** (go to question 11)
☐ **yes.** If so, could you then give an estimate (in costs and/or in man years) of the volume of your external R&D?

Volume of external R&D in 1988: (please give man years and/or costs)
Dutch guilders: . . . .
man years: . . . .

Notes

1. Among the "soft signals" that such a bias might exist are the following: first, smaller firms which more frequently than larger firms report having no R&D, had a lower response rate; second, inspection of the questionnaires suggests that highly innovative firms filled in their questionnaires more carefully than did firms with little innovation activity; third, we received a number of phone calls in which firms wondered whether it was really useful to participate in the survey, as they did not have any innovation activities (in spite of our emphasis, in the accompanying letter, that non-innovative firms should also participate).
2. The "principal enterprise establishment" (in Dutch *hoofdvestiging*) is often but not always identical with the headquarters of a firm. However, in most cases it is more than a mere production site, as it may have considerable management power.
3. In an analysis of the effects of US tax credits for R&D, Cordes (1989) found indications of such a redefinition of R&D.
4. Some measurement differences may also result from different observation periods. The subsidy records relate to the period from October 1987 to September 1988, the most recent R&D data from the CBS relate to 1985 (for firms having 50 and more employees) and 1986 (for firms having 10–49 employees), whereas the SEO survey relates to 1988. We think that such differences do not matter much when comparing numbers of firms having R&D, as this number remained almost constant during the period 1983–88 according to our two surveys (Kleinknecht *et al.*, 1990). However, it makes a difference when looking at man years of R&D, as the latter have strongly increased in recent years.
5. In an analogy, this could be compared to collecting investment data under the restriction that investment projects beyond a certain minimum amount should be left out. While this would be convenient for the data collectors, subsequent analyses of these investment data could lead to problematic conclusions. For example, one might conclude from such investment data that small firms are less capital intensive than larger ones.

References

Baldwin, W. L. and J. T. Scott (1987) *Market Structure and Technological Change*, London: Harwood Academic Publishers.
CBS (1985) *Research and Development in the Netherlands in 1985* (in Dutch), The Hague: Central Statistical Office.

CBS (1986) *Research and Development in Enterprises having less than 50 Employees in 1986* (in Dutch), The Hague: Central Statistical Office.

Cordes, J. J. (1989) "Tax incentives and R&D spending: A review of the evidence" in *Research Policy*, vol. 18, pp. 119–33.

Hughes, K. (1986) *Exports and Technology*, Cambridge: Cambridge University Press.

Kamien, M. I. and N. L. Schwartz (1982) *Market Structure and Innovation*, Cambridge: Cambridge University Press.

Kleinknecht, A. (1987a) "Measuring R&D in small firms: How much are we missing?" in *Journal of Industrial Economics*, vol. 36, December, pp. 253–6.

Kleinknecht, A. (1987b) *Innovation Patterns in Crisis and Prosperity*, London: Macmillan; New York: St Martin's Press.

Kleinknecht, A. and B. Verspagen (1989) "R&D and market structure: The impact of measurement and aggregation problems" in *Small Business Economics*, vol. 1, December, pp. 297–301.

Kleinknecht, A., J. O. N. Reijnen and J. J. Verweij (with the assistance of Joost Hagens and Ineke Roest) (1990) *Innovation in the Dutch Manufacturing and Service Sectors* (in Dutch), The Hague: Ministry of Economic Affairs (Beleidsstudies Technologie-Economie).

Levin, R. C., W. M. Cohen and D. C. Mowery (1985) "R&D appropriability, opportunity and market structure: New evidence on some Schumpeterian hypotheses" in *American Economic Review, Papers and Proceedings*, vol. 75, no. 2, pp. 20–4.

Scherer, F. M. (1990) *Industrial Market Structure and Economic Performance*, 3rd edition, Boston: Houghton Mifflin.

Scott, J. T. (1984) "Firm versus industry variability in R&D intensity" in Z. Griliches (ed.) *R&D, Patents and Productivity*, Chicago, IL: University of Chicago Press, pp. 233–45.

7

A Poisson model of patenting and firm structure in Germany*

Joachim Schwalbach and Klaus F. Zimmermann

7.1 Introduction

Firms invest in productive resources only if they expect a positive return on the investment. The opportunity costs to firms, however, vary between the types of investments, depending upon the level of risk associated with them. Investments in R&D are generally characterized by a higher level of riskiness than, for instance, investments in product imitation or advertising and, therefore, the expected future profit rates should be higher. As a consequence, firms only commit resources for R&D if expected return exceeds that earned on other kinds of alternative investment. The amount of resources committed to R&D increases as long as the marginal rate of return is positive. Resulting positive profits might attract new firms to the market, forcing incumbent firms to commit R&D efforts at a higher level than the profit maximization level, if incumbents still expect to maintain their competitive advantage.

A host of studies exist which analyze the inter-relationship between technological investment, firm structure and market characteristics. Many empirical studies concentrated on the Schumpeterian hypothesis that market structure affects the technological outcome.[1] Other empirical studies investigated R&D production functions which transform R&D inputs into inventions, reflecting also the randomness of the R&D

* Financial support from the German Science Foundation (DFG) in the form of a Heisenberg Fellowship to the second author is gratefully acknowledged. We are grateful to William Comanor, Paul Geroski and Klaus Herdzina for very valuable comments and suggestions. We wish to thank Martin Hofmann, John De New, Stefan Csutor and Sabine Hetebrüg for able research assistance.

process.[2] Theoretical studies, on the other hand, concentrated on decision models which deal with optimal scale and timing of R&D investments, as well as life time of patents and copyrights. Another branch of theoretical work considers R&D efforts in a game-theoretic context in which the players try to increase their probability of winning the R&D race.[3]

In this chapter, we adopt the investment theoretic approach to R&D decisions and examine the hypothesis of whether firms' R&D efforts depend on current and expected post-invention profit rates. The hypothesis is tested for the first time for Germany by considering a sample of 143 German corporations.

Although many studies have shown that R&D efforts and profit rates are positively related, only a few take the lag structure between the two factors into account. Among them the studies by Branch (1974), Elliott (1971), Geroski (1987), and Ravenscraft and Scherer (1982).[4] Branch estimated distributed lag functions for a pooled time series of patents and profit rates for 111 US corporations for the time period 1950–65. He found that profit rates increase with increasing R&D success and also that high profit rates stimulate R&D efforts. In Elliott's study, the relationship between R&D efforts and profitability is less clear, which leads Elliott to the conclusion that R&D efforts are far more stimulated by the extra-profit expectation than by the actual profit rates realized. Geroski used panel data on 4,378 significant innovations introduced in the United Kingdom between 1945 and 1983 in 73 three-digit industries. He found that post-innovation profit rates are positively related to the number of innovations but to a diminishing degree. Ravenscraft and Scherer used the PIMS time series data of the Strategic Planning Institute and found that the mean lag between R&D inputs and profit returns ranged between four to six years for the time period 1970–79. In addition, they showed that profit returns on R&D were higher in the late 1970s than during the early 1970s, but the general positive relationship between R&D and profitability persisted over time.

The chapter is organized as follows: in Section 7.2 we present the theoretical model. Section 7.3 describes the data sample, Section 7.4 reports on the empirical model, and Section 7.5 provides the empirical results. Section 7.6 concludes the chapter.

7.2   The theoretical framework

Investment theory suggests that firms invest in R&D activities if the expected post-invention rate of return exceeds the return that would be earned otherwise by an amount sufficient to justify the investment.

Further, the more resources those firms invest, the higher are the required profit rates. The difficulty is measuring the return that would be earned in the absence of the investment. For this purpose, we let R&D expenditures be a function of the difference between expected post-invention and current profit rates, and also the difference between a firm's current and the (economy-wide) average profit rate. These relationships can be formally specified as follows:

$$R\&D_{it} = f(\Pi^*_{it+1} - \Pi_{it'}\Pi_{it} - \Pi_{t'}\psi^{(1)}_{it}) \tag{7.1}$$

where:

$R\&D_{it}$ is R&D inputs by firm i in period t;
Π^*_{it+1} is expected post-invention profit rates;
Π_{it} is current profit rates;
Π_t is average profit rate of the economy; and
$\psi^{(1)}_{it}$ represents all other factors affecting R&D activities.

Equation 7.1 shows that firm i commits more resources to R&D in period t if $\Pi^*_{it+1} - \Pi_{it}$ or $\Pi_{it} - \Pi_t$ is higher. However, R&D investment decisions are subject to other influences which are reflected by $\psi^{(1)}_{it}$.[5] These determinants are discussed later. A convenient functional specification for Equation 7.1 is

$$R\&D_{it} = \alpha_0 + \alpha_1(\Pi^*_{it+1} - \Pi_{it}) + \alpha_2(\Pi_{it} - \Pi_t) + \alpha_3\Psi^{(1)}_{it} \tag{7.2}$$

with $\alpha_1, \alpha_2 > 0$.

The expected positive impact of R&D on a firm's success or profits implies that the firm is able to transform R&D inputs into inventions and after all into patents, licensing and/or innovations. This process is stochastic. In this study we focus on the number of patents as the only measure of the R&D process available to us.[6]

Since patents are a discrete variable, a natural choice for a data generating framework is to assume a Poisson process. The Poisson distribution is widely accepted as one reasonable econometric approach for analyzing events which occur both randomly and independently in time.[7]

The basic Poisson probability specification is

$$\Pr(P_{it+1}) = \frac{\exp(-\mu_{it+1})\mu_{it+1}^{P_{it+1}}}{P_{it+1}!} \tag{7.3}$$

where P_{it+1} is the observed event count for firm i during the time period t and μ_{it+1} is a time-dependent and individual-specific parameter which equals the conditional expectation of

$$P_{it+1} : E(P_{it+1}|\cdot) = \mu_{it+1}$$

A further advantage of this specification is that $P_{it+1} = 0$ is a natural outcome of the Poisson specification. The deterministic part of Equation 7.3, μ_{it+1}, is specified such that

$$\log \mu_{it+1} = \beta_0 + \beta_1 R\&D_{it} + \beta_2 \psi_{it}^{(2)} \tag{7.4}$$

whereas the stochastic part of the model comes from the Poisson specification in Equation 7.3. $\beta_1 > 0$ and $\psi_{it}^{(2)}$ represents all other factors affecting the patenting activity. These determinants will be discussed below.

By combining Equations 7.2 and 7.4, we obtain

$$\log \mu_{it+1} = \beta_0 + \beta_1[\alpha_0 + \alpha_1(\Pi_{it+1}^* - \Pi_{it}) \tag{7.4'}$$
$$+ \alpha_2(\Pi_{it} - \Pi_t) + \alpha_3 \psi_{it}^{(1)}] + \beta_2 \psi_{it}^{(2)}$$

Equations 7.3 and 7.4' form our basic behavioral framework which can be estimated by a maximum likelihood algorithm which is an easy task given that the log–likelihood function is globally concave. Note, however, that we are not able to identify α_1, α_2 and β_1 separately, and that for factors common to $\psi_{it}^{(1)}$ and $\psi_{it}^{(2)}$ only the joint impact will be measured.

7.3   The basic data sample

The analysis is made possible by combining two uniquely rich data sets of the same 143 German manufacturing companies: a data set on firm-specific patent activity and a data set on financial and other information. The data set on patenting consists of the stock of patents each of the 143 companies held in 1982. We use the patent data as a measure of R&D success. Patent information was collected for all companies from the German Patent Office files. Access to these files was limited[8] to the extent that only the total number of patents issued to the companies could be made available. Patents are often not issued to the companies themselves but instead to other firms owned by these companies. As a consequence, we had to collect the patent record by these firms as well. As a result, we know the number of patents directly issued to the 143 companies and/or to those firms owned by the company.

Table 7.1 summarizes the frequency distribution and descriptive statistics of the patent data. Column 1 provides the statistics for patents issued to the parent company itself, Column 2 for the subsidiaries, and Column 3 for the total number of companies' patents. Column 1 shows

Table 7.1 Frequency distribution and descriptive statistics of patents by 143 German corporations.

	Number of companies in patent size class		
Number of patents	Parent company (1)	Subsidiary (2)	Parent company and subsidiaries (3)
0	43	99	40
1–10	36	12	30
11–100	33	18	35
101–500	18	7	21
501–1 000	6	2	6
>1 000	7	5	11
Descriptive statistics			
Maximum	7 390	5 346	9 805
Mean	193.10	120.29	313.39
STD	743.06	572.61	1 100.30

that for 43 companies, no patent has been issued in twenty years (1963–82), while for the remaining companies the number of patents varies greatly. In addition, one has to take into account the patents which are held by the subsidiaries. Column 2 illustrates that for 99 companies, none of the subsidiaries held a positive patent record, and the remaining companies decentralized their R&D activities. Column 3 represents the summary statistics for the total number of companies' patents. According to Column 3, one finds 40 out of the 143 companies for which no patents are reported. We conjecture that most of these companies do not invest greatly in R&D activities, and the others may operate R&D divisions but did not succeed in obtaining a patentable invention or are members of industries which have a record not to seek patent protection. We also do not want to exclude the possibility that some companies received patents but did not invest into R&D in any significant way.[9] The patent data set also reveals that in three companies, for which patentable inventions were reported, R&D activities were concentrated outside the company in subsidiaries. However, R&D oriented companies decentralized their activities in the sense that R&D success is reported for the company as well as for the outside research facilities.

The patent data set is combined with data on financial information which was gathered from balance sheets and business reports of the 143 companies. For these companies we could rely on a consistent time series of financial data for the years 1966 to 1985. In addition, we had access to information on the multi-industry activities of these companies. Industry-specific data were collected from published sources of the Statistische Bundesamt in Wiesbaden.

7.4 Empirical model and basic hypotheses

The empirical model used in this section follows the theoretical framework outlined in Section 7.2, especially in Equations 7.3 and 7.4'. As became apparent in the previous section, we have only one cross-section available. R&D activities are measured by the number of patents held by each of the German companies in the year 1982. However, we are also able to identify the firm's main market so as to connect the firm data with the associated industry data on the two-digit level. We can, therefore, identify industry effects and control for heteroskedasticity in the data.

The modified empirical framework is the Poisson probability specification

$$\Pr(P_{ij}) = \frac{\exp(-\mu_{ij})\mu_{ij}^{P_{ij}}}{P_{ij}!} \tag{7.3'}$$

and a reformulation of Equation 7.4':

$$\log \mu_{ij} = \gamma_0 + \gamma_1(\Pi_{ij}^* - \Pi_{ij}) + \gamma_2\Pi_{ij} + \gamma_3\psi_{ij} \tag{7.4''}$$

Note that $\gamma_0 = \beta_0 + \beta_1(\alpha_0 - \alpha_2\Pi)$, $\gamma_1 = \beta_1\alpha_1$, $\gamma_2 = \beta_1\alpha_2$, ψ_{ij} contains all elements of $\psi_{ij}^{(1)}$ and $\psi_{ij}^{(2)}$, and γ_3 is the vector of joint corresponding parameters. The vector ψ_{ij} consists of firm-specific and industry-specific variables. In the empirical analysis, we include the following variables:

S_{ij}: firm size measured by the yearly average of sales in 1966–1975;
S_{ij}^2: firm size squared;
C_{ij}: capital intensity calculated as yearly averages in 1966–1975;
D_{ij}: market diversification as measured by the number of four-digit industries the firm supplied in 1980; and
R_{ij}: market risk determined by the covariance of the firm-specific profit rate and the economy profit rate for the period 1966–1975.

The industry variables are two (1,0)-dummies for consumption goods (CD_i) and investment goods (ID_j) leaving all other industries as the reference group. We also include the industry export rate EX_j and the industry import rate IM_j, where both are measured at the two-digit level, P_j the average number of patents in the industry calculated from the sample at hand at the two-digit level, and H_j the Herfindahl index as a measure of market concentration.

Finally, note that the expected post-patent profits, Π_{ij}^*, are not directly observable, and the usual procedure that has been followed is to proxy it by lagged actual profits. As the appropriate time-lag, we choose ten years which is the average life time of German patents.[10] Therefore,

Π_{ij}^* and Π_{ij} are the actual average yearly profit rates for 1976–1985 and 1966–1975, respectively. According to our initial framework, both Π_{ij}^* and Π_{ij} exhibit a positive impact on the incentive to invest in R&D and, given that $\beta_1 > 0$, $\gamma_1, \gamma_2 > 0$.

The Schumpeterian hypothesis that large firms are more conducive to technological progress than smaller firms has been tested extensively. Others have reported diminishing marginal returns in the R&D output–input relationship.[11] Recent studies even have shown a U-shaped relationship between firm size and innovative activities.[12] Taking these results into account, we test this hypothesis by estimating a quadratic relationship between patenting and firm size. This form permits us to examine whether an optimum firm size is present.

Related to the Schumpeterian hypothesis is the conjecture that large firms operate in various markets simultaneously because their R&D activities enhance the cross-fertilization of technological knowledge. Most empirical studies have found statistical support for R&D-based diversification into related product markets.[13] Thus, we expect $\gamma_{34} > 0$.

Technological opportunities and the propensity to patent vary considerably across industries and are most pronounced in industries such as pharmaceuticals, chemicals, machinery/mechanical engineering and electrical equipment.[14] Most firms in these industries operate at a higher level of capital intensity than firms with lower technological opportunities. Capital intensity might, therefore, be a good substitute for the technological potential of the firm and its related markets. Thus, we expect $\gamma_{33} > 0$.

R&D investments are subject to high risk. However, inventive activity can be one strategy to reduce the dependence of profits on general business conditions. The sign of the parameter of market risk R_{ij}, γ_{35}, is, therefore, ambiguous.

Various industry measures further control for sample heterogeneity. A general impact is reflected in the parameters of the dummies for consumption and investment goods, CD_j and ID_j, respectively. We expect that consumption goods industries have a smaller and investment goods industries a larger incentive to patent than the reference group. Hence, $\gamma_{40} < 0$ and $\gamma_{41} > 0$.

Export and import rates (EX_j and IM_j) measure the extent to which firms are forced to compete by innovative activity. However, the larger the export rate, the more likely that patent activity will lead to patent applications outside Germany. Thus, we expect $\gamma_{36} \gtrless 0$ and $\gamma_{37} > 0$. A further measure of technological opportunities is P_j, the average number of patents held in the industry. This suggests $\gamma_{38} > 0$. Finally, $\gamma_{39} > 0$ would further indicate that market power, measured by the Herfindahl index H_j, spurs inventive activity.

7.5 Empirical results

The model outlined in the last section consisting of Equations 7.3', 7.4"
and 7.5 is estimated using maximum likelihood procedures and
Newton's Method. The richness of our data enables us to analyze the
patent data from the parent company separately from those held by its
subsidiaries. From a theoretical standpoint, the number of patents held
by the company as a whole is the more satisfactory approach because
subsidiaries are often held simply for their innovative potential, whereas
patents are exploited throughout the entire company. Parameter
estimates for the complete model are given in Table 7.2 together with
the asymptotic standard errors, a χ^2-statistic and elasticities calculated
from the model results.

Table 7.2 Poisson models of patenting activity.[a]

Variables	Parent company		Parent company and subsidiaries	
	Estimates	Elasticities	Estimates	Elasticities
Constant	3.000	—	3.619	—
	(0.039)		(0.031)	
CD	−2.087	—	−2.460	—
	(0.061)		(0.055)	
ID	−0.282	—	0.376	—
	(0.019)		(0.015)	
$\Pi^* - \Pi$	1.522	−0.048	1.725	−0.054
	(0.172)		(0.144)	
Π	6.786	0.447	7.480	0.492
	(0.212)		(0.192)	
$S \times 10^{-6}$	0.852	0.481	0.868	0.490
	(0.011)		(0.009)	
$(S \times 10^{-6})^2$	−0.053	−0.149	−0.058	−0.163
	(0.001)		(0.001)	
C	0.128	0.103	0.262	0.212
	(0.011)		(0.007)	
D	0.027	0.215	0.034	0.271
	(0.001)		(0.001)	
R	−0.257	0.380×10^{-8}	0.094	0.225×10^{-7}
	(0.012)		(0.009)	
$EX \times 10^2$	0.020	0.415	−0.013	−0.271
	(0.001)		(0.001)	
IM	−0.312	−0.051	0.543	0.089
	(0.102)		(0.081)	
$P \times 10^{-3}$	0.236	0.037	0.388	0.061
	(0.038)		(0.036)	
$H \times 10^{-3}$	0.483	0.034	−2.122	−0.147
	(0.168)		(0.169)	
χ^2	23 053.0		32 936.0	

Note: [a]The endogenous variable is the number of patents. Standard errors in parentheses.
All estimates are significant at the 5 per cent level. Elasticities are calculated at the sample
means. Sample size is 143.
$\chi^2 = \sum_i (P_{ij} - \mu_{ij})^2 / \mu_{ij}$

To interpret these estimates, note that the conditional expectation of the number of patents per firm is

$$E(P_{ij}) = \mu_{ij} \tag{7.5}$$

Assume

$$\log \mu_{ij} = \delta_0 + \delta_1 Z_{ij} + \delta_2 Z_{ij}^2 \tag{7.6}$$

Then it is straightforward to show that

$$\frac{\partial \log E(P_{ij})}{\partial Z_{ij}} = (\delta_1 + 2\,\delta_2 Z_{ij})E(P_{ij}) \tag{7.7}$$

Hence,

$$\varepsilon(E(P_{ij}), Z_{ij}) = \delta_1 Z_{ij} + 2\delta_2 Z_{ij}^2 \tag{7.8}$$

where $\varepsilon(\cdot)$ stands for the elasticity.

Comparing the findings for patents of the parent company and for the parent company and subsidiaries together, one finds that parameters change sign for the investment goods dummy (ID), the proxy of risk (R), and the export and import rates $(EX$ and $IM)$. It is therefore crucial to choose the relevant concept.

Results for the joint patents of parent company and subsidiaries are very consistent with the theoretical expectations outlined in the last section: patent activity increases if the difference between the expected post-patent rates of return and current profit rates increases. More relevant, however, seems to be that realized profits in $t - 1$ increase the likelihood of patents in t. The parameter of Π is much larger than that of $(\Pi^* - \Pi)$ and the elasticity is 0.492.

Results for firm size (S) indicate that $\gamma_{31} > 0$ and $\gamma_{32} < 0$. Therefore, we observe a maximum in the relevant range with first positive but diminishing marginal returns and then negative marginal returns. The joint elasticities of S and S^2 are 0.183 for the parent company only and 0.164 for the joint patents of the parent company and the subsidiaries. The capital–labor ratio (C) and the index of diversification (D) both exhibit the expected positive parameter estimates and the elasticities are 0.212 and 0.271, respectively. Patents are positively influenced by the measure of risk (R), although the elasticity is of negligible size.

The parameter estimate of the export rate (EX) is negative indicating that German patent activity addressed to the German Patent Office is smaller *ceteris paribus* in internationally oriented industries, and the elasticity is large in absolute terms (0.271). Import competition measured by IM, however, has a positive impact on the number of patents with an elasticity of 0.089. Technological opportunities (P) have the expected positive parameter with elasticity 0.061. Finally, the coefficient of the Herfindahl index (H) indicates a negative effect of market power, with an elasticity -0.147; this result is contrary to the Schumpeterian hypothesis.

7.6 Conclusions

The access to a unique data set enabled us to test for Germany for the first time the hypothesis that firms only invest in R&D activities if they can expect that post-invention rate of return exceeds current profit rates. The sample consists of 143 German corporations for which the stock of patents held by these firms in 1982 is known.

The empirical results are in line with most of the studies mentioned at the beginning. The results show that R&D investments lead to a positive profit difference between post- and pre-invention rate of return and contribute to a stabilization of above average profit rates. Furthermore, the relationship between firm size and patenting is inverse U-shaped, and capital-intensive, risk-taking and diversified firms show a higher extent of patenting. In addition, the results show that firms in export-oriented industries are less inclined to patenting than in import-oriented industries. The results also stress the positive relationship between technical opportunity and the propensity to patent as well as the negative relationship between a seller concentration and patent yield.

Data Appendix

Variables	Definition
P_{ij}	Number of patents firm i held in 1982
Π_{ij}^*	Yearly average of profit rates, 1976–1985
Π_{ij}	Yearly average of profit rates, 1966–1975
S_{ij}	Size of firm i, measured as the average sales within the period 1966–1975
D_{ij}	Extent of market diversification of firm i, measured by the number of four-digit industries firm i was operating in 1980
C_{ij}	Capital intensity of firm i, measured by the ratio of total assets to sales within the period 1966–1975
R_{ij}	Risk of firm i, measured as the covariance between firm's profit rates and economy-wide profit rates within the period 1966–1975
CD_{ij}	(0,1)-dummy for consumption goods industries
ID_{ij}	(0,1)-dummy for investment goods industries
EX_j	Export rate, in per cent, 1966–1975, two-digit level
IM_j	Import rate, in per cent, 1966–1975, two-digit level
P_j	Average number of patents held in 1982 in a two-digit industry, calculated from the sample at hand
H_j	Herfindahl index 1977

Notes

1. For a survey of this literature, see Scherer (1980), Kamien and Schwartz (1982), Baldwin and Scott (1987) and Dosi (1988).
2. See for instance, for an earlier study Comanor (1965) and for more recent ones the studies in Griliches (1984).
3. For a survey of the literature, see Kamien and Schwartz (1982) and for a more recent study, see Dixit (1987).
4. For other studies, see Grabowski and Mueller (1978).
5. Equation 7.1 represents a type of model which can be adapted to other kinds of investment decisions. For instance, it has been used for investments in new markets, see Schwalbach (1987).
6. At this point, we do not want to repeat the discussion of how good patents are as a measure of the R&D process and whether patents are a measure of R&D output or input. For details see Comanor and Scherer (1969) and Basberg (1987).
7. See Hausman *et al.* (1984).
8. We would like to express our gratitude to the President of the German Patent Office, Dr Erich Häuser, for supporting the project and to Mrs Müller and Mr Rausch for their great help in collecting the patents.
9. For empirical evidence that some firms received patents without reporting any R&D effort, see Scherer (1983).
10. The average life time of a patent in 1988 was twelve years and eleven months, beginning with the patent registration. On average, it takes slightly more than two years until a patent is granted. For more details, see Deutsches Patentamt, *Jahresbericht*, 1988.
11. See for instance Scherer (1983) for the United States and Zimmermann (1987, 1989) for Germany.
12. See for instance Pavitt *et al.* (1987).
13. See for instance Pavitt *et al.* (1987), Scherer (1983) and Schwalbach (1989).
14. For empirical evidence see Mansfield (1986), Pavitt *et al.* (1987) and Scherer (1983).

References

Baldwin, William L. and Scott, John T. (1987), *Market Structure and Technological Change*, London: Harwood Academic Publisher.
Basberg, Bjorn L. (1987), "Patents and the measurement of technological change: A survey of the literature", *Research Policy*, vol. 16, no. 2–4, August, pp. 131–41.
Branch, Ben S. (1974), "Research and development activity and profitability: A distributed lag analysis", *Journal of Industrial Economics*, vol. 82, no. 4, September–October, pp. 999–1011.
Comanor, William S. (1965), "Research and technical change in the pharmaceutical industry", *Review of Economics and Statistics*, vol. 47, May, pp. 182–91.
Comanor, William S. and Scherer, Frederic M. (1969), "Patent statistics as a measure of technical change", *Journal of Political Economy*, vol. 77, May–June, pp. 392–8.

Deutsches Patentamt, *Jahresbericht*, 1988.

Dixit, Avinash (1987), "Strategic behavior in contests", *American Economic Review*, vol. 77, no. 5, December, pp. 891–8.

Dosi, Giovanni (1988), "Sources, procedures and microeconomic effects of innovation", *Journal of Economic Literature*, vol. 26, no. 3, September, pp. 1120–71.

Elliott, Walter J. (1971), "Funds flow vs. expectational theories of research and development expenditures in the firm", *Southern Economic Journal*, vol. 37, April, pp. 409–22.

Geroski, Paul (1987), *Innovation, Technological Opportunity and Market Structure*, mimeo, September.

Grabowski, Henry G. and Mueller, Dennis C. (1978), "Industrial research and development, intangible capital stocks and firm profit rates", *Bell Journal of Economics*, vol. 9, pp. 328–43.

Griliches, Zvi (ed.) (1984), *R&D, Patents and Productivity*, Chicago: Chicago University Press.

Hausman, Jerry, Hall, Bronwyn H. and Griliches, Zvi (1984), "Econometric models for count data with an application to the patents–R&D relationship", *Econometrica*, vol. 52, no. 4, July, pp. 909–38.

Kamien, Morton I. and Schwartz, Nancy L. (1982), *Market Structure and Innovation*, Cambridge: Cambridge University Press.

Mansfield, Edwin (1986), "Patents and innovation: An empirical study", *Management Science*, vol. 32, no. 2, February, pp. 173–81.

Pavitt, Keith, Robson, Michael and Townsend, Joe (1987), "The size distribution of innovating firms in the UK: 1945–1983", *Journal of Industrial Economics*, vol. 35, no. 3, March, pp. 297–316.

Ravenscraft, David and Scherer, Frederic M. (1982), "The lag structure of returns to research and development", *Applied Economics*, vol. 14, December, pp. 603–20.

Scherer, F. M. (1980), *Industrial Market Structure and Economic Performance*, Boston: Houghton Mifflin Company.

Scherer, F. M. (1983), "The propensity to patent", *International Journal of Industrial Organization*, vol. 1, no. 1, March, pp. 107–28.

Schwalbach, Joachim (1987), "Entry by diversified firms into German industries", *International Journal of Industrial Organization*, vol. 5, no. 1, March, pp. 43–9.

Schwalbach, Joachim (1989), "Small business in German manufacturing", *Small Business Economics*, vol. 1, no. 2, May, pp. 129–36.

Schwalbach, Joachim (1990), *Diversifizierung, Risiko und Erfolg industrieller Unternehmen*, Tübingen: Mohr & Siebeck.

Zimmermann, Klaus F. (1987), "Trade and dynamic efficiency", *Kyklos*, vol. 40, no. 1, pp. 73–87.

Zimmermann, Klaus F. (1989), *Innovative Activity, Employment Decisions and the Neoclassical Model of the Firm: Theory and Practice*, Mannheim, unpublished manuscript.

8

Technological diffusion, firm size and market structure

Paul Stoneman

8.1 Introduction

In this chapter I explore in a relatively simple way the relationship between the diffusion of new technology, firm size and market structure. There are several reasons for looking at the diffusion stage rather than the R&D stage. First, I feel that the R&D literature has been pretty well surveyed and explored elsewhere. Second, I am much more interested in diffusion than R&D and much more attuned to what is happening in the diffusion field. Third, and most importantly, I tend to feel that diffusion is the poor relation in the technological change literature. It has merited much less attention than R&D, although in reality it is only as diffusion proceeds and new technologies are used that such new technologies have their impact and their benefits are realized. Finally, I do not actually know of a large body of work that relates diffusion to market structure and thus exploring this particular issue is more exciting than looking once again at R&D.

For the purposes of this chapter I will take market structure as shorthand for the size distribution of firms. Clearly there are many dimensions of market structure that this does not cover, e.g. barriers to entry, but as a focal point this definition is useful. Throughout the chapter I will use "concentration" as synonymous with market structure. The concept of firm size is self-explanatory.

Throughout the chapter I have in mind an economy that can be represented by a capital goods-supplying sector and a capital goods-using sector that provides final products. New technologies embodied in capital goods are assumed to come from the capital goods sector, are used in the capital-using sector and are here called process innovations.

New consumer products are assumed to originate in the capital goods-using sector and are called product innovations.

With this conception of the economy, market structure will refer to the size distribution of firms in the capital goods-producing and the capital goods-using sectors. The relation between diffusion and market structure is thus a relation between diffusion and the size distributions of firms in the two sectors. It goes almost without saying, that the relation between diffusion and market structure is a two-way relationship, market structure affects diffusion patterns and diffusion patterns affect market structure. It is the same in the R&D literature. However, by looking at diffusion we can gain an extra dimension that is more dynamic than we are used to. If diffusion affects market structure, then at different stages of the diffusion process we may expect to see different market structures in the industry into which the new technology is being diffused. We thus explore the relation between the diffusion path and the *path* of market structure. Moreover, if, as we shall argue, firm size is endogenous, then the relation of diffusion to firm size will not be a simple one.

The work reported upon in this chapter is theoretical rather than empirical. The chapter proceeds by the consideration of relationships between diffusion, firm size and market structure in several simple alternative modeling frameworks. The emphasis is on models of inter-firm diffusion, rather than intra-firm diffusion. Thus the questions being asked concern how market structure and the inter-firm diffusion path are related and whether there is any relationship between firm size and the date of adoption of new technology. To be specific we begin with the investigation of a simple probit diffusion model, concentrating on process innovations, in which the supply of the new technology being diffused simply accommodates the demand for that technology, and in which market structure and firm sizes in the using industry are not affected by diffusion. Next we introduce the supply side by concentrating again on a new process technology, and again assuming that the using industry market structure is not affected by diffusion. Third, we once more assume an accommodating supply industry, but allow now that diffusion does affect market structure and firm sizes in the using industry and see what patterns emerge. Finally, we draw some conclusions.

8.2   Market structure, firm size and the diffusion of a new process technology in a demand-based probit model

Davies (1979) finds, assuming other determining factors fixed, that the rate of diffusion of a new process technology is, empirically, negatively related to the number of firms (N) in the using industry and the variance of the log of firm size (σ_x^2) in that industry. Although most concentration measures can be considered as weighted functions of N and σ_x^2, until the weights are specified it is not possible to say whether, from these results, diffusion speed increases or decreases with concentration. Davies' results were derived from a probit framework. Romeo (1977), working in an epidemic/logistic framework, found a positive coefficient on σ_x^2, but it is not really possible to explain that finding with any degree of precision in a logistic framework.

In Davies' (1979) probit framework which is very similar to that of David (1969) the findings are not too difficult to rationalize. It is assumed that there is a new process technology that a firm may acquire by the purchase of one (and only one) unit of a new capital good at price p_t in time t. The returns to ownership (R) vary with firm size (S) positively, say $R = \gamma S$. Let S be distributed according to the cumulative density function F(S). Then a firm will acquire a new unit of capital in time t if p_t, the cost of acquiring the technology in time t, is such that

$$p_t \leqslant R = \gamma S \tag{8.1}$$

or if

$$S \geqslant \frac{p_t}{\gamma} \tag{8.2}$$

The number of owners at time t (M_t) is then

$$M_t = N\left(1 - F\left(\frac{p_t}{\gamma}\right)\right) \tag{8.3}$$

Take a very simple distribution for F(S), say the logistic distribution, then

$$F(S) = \frac{1}{1 + \exp[-(S - \alpha)/\beta]} \tag{8.4}$$

where mean (S) = α and var(S) = $\beta^2\pi^2/3$. From Equations 8.3 and 8.4

$$\log_e\left(\frac{M}{N - M}\right) = \frac{\alpha}{\beta} - \frac{p_t}{\beta} \tag{8.5}$$

Thus for a given price, p_t, we see that

$$\frac{\partial M}{\partial N} = \frac{M}{N} > 0 \tag{8.6}$$

$$\frac{\partial M}{\partial \alpha} = \frac{1}{\beta} \cdot \frac{M}{M(N - M)} > 0 \tag{8.7}$$

and

$$\frac{\partial M}{\partial \beta} = \frac{1}{\beta} \left[-\log\left(\frac{M}{N - M}\right) \right] \tag{8.8}$$

If $M/(N - M) < 1$, $\log[M/(N - M)] < 0$. Thus if $M < (1/2)N$, $(\partial M/\partial \beta) > 0$ and if $M > (1/2)N$, $(\partial M/\partial \beta) < 0$. Given these results an increase in N increases use, but the elasticity of M with respect to N is unity. An increase in the mean of firm size, α, also increases use. An increase in the variance of firm size, β, increases use for $M < (1/2)N$, and decreases use for $M > (1/2)N$.

These results on market structure, obviously, do not match the empirical results of Davies but one may see the way that they are generated from a simple model. Any findings with respect to firm size in this framework are really incorporated by assumption. If the returns increase with firm size then obviously large firms adopt earlier. Moreover, by assumption, in the model, if two small firms merge to make one large firm, the date of adoption by the merged unit would be earlier than that of the two smaller units.

As far as diffusion models go, this is very simple. By making the return a function of firm size, it gives a direct link between diffusion and the firm size distribution. However, this is not essential. One can construct models in which firms differ across some other dimension and allow the F( ) distribution to be a distribution of "reservation prices" instead. One can then argue that the mean and/or variance of these reservation prices is related to some measure of market structure and get a similar result.

There are, however, more interesting paths to pursue, so we will forgo that exercise. One other problem with this probit model is the technique choice rule, $p_t < R$. In a world where p_t is falling, as it must to drive the diffusion (if R does not increase with t), this may be an inappropriate rule. It states that the technology is adopted at the first date it is profitable to adopt. However, the actual rule should be that adoption occurs at that date when it is *most* profitable to adopt. This rule will involve the potential buyers' expectations on p_t. If buyers are myopic then $p_t < R$ is the rule that results. However, in the absence of myopia, the expectation of p_t will enter the decision rule. To fully explore this however, we need to add a supply side.

8.3   The supply side

In a series of papers with Norman Ireland (e.g. Ireland and Stoneman, 1986), I have explored diffusion models of the probit type in which a supply side is explicitly incorporated. We will concentrate on that detailed in Ireland and Stoneman (1986).

As before, buyers only purchase one unit of a new capital good. We index buyers in decreasing order of the service flow, such that a buyer with index x obtains a constant flow of services $g(x)$ per period of ownership and $g_x < 0$. For the xth ranked buyer, two conditions have to be satisfied for adoption in time t.

$$-p(t) + \frac{g(x)}{r} \geq 0 \qquad \text{(profitability condition)} \qquad \textbf{(8.9)}$$

$$-D\hat{p}(t) + rp(t) - g(x) \leq 0 \qquad \text{(arbitrage condition)} \quad \textbf{(8.10)}$$

where $D\hat{p}(t)$ is a continuous time representation of the buyers' expectation of the change in price between t and $t + dt$. (We use the D operator to represent a derivative with respect to time.) Under myopia $D\hat{p}(t) = 0$, and thus the xth indexed buyer will acquire at the first t when $rp(t) = g(x)$. Under perfect foresight, the only other case considered is $D\hat{p}(t) = Dp(t)$, and given, as is the case, $Dp(t) < 0$, the arbitrage condition dominates the profitability condition and Equation 8.10 defines the dynamic demand for the technology. We note that as buyers only ever buy one unit, the rank x of the marginal buyer in time t equals the number of acquirers to time t. If the ranking of buyers is related to firm size, again we get the result that large firms are the first to acquire.

On the supply side we assume n quantity setting, identical firms, each maximizing its expected profit given the behavior of the other $(n - 1)$ firms. We assume that while production takes place unit cost $c(t)$ falls until some time $\hat{t}$, after which it increases, i.e.

$$Dc(t) \lessgtr 0 \qquad \text{as} \qquad t \lessgtr \hat{t} \qquad \textbf{(8.11)}$$

It is then shown that on the optimal profit maximizing path, the two trajectories under myopia and perfect foresight can be characterized by:

Myopia:

$$rp = g(x) = rc - Dc + \frac{n - 1}{n} \cdot Q \frac{g_x}{r} \qquad \textbf{(8.12)}$$

$$g(x_1) = rc(t_1) \qquad \textbf{(8.13)}$$

Perfect foresight:

$$-Dp + rp = g(x) = rc - Dc - \frac{xg_x}{n} \qquad \textbf{(8.14)}$$

$$g(x_1) = rc(t_1) - \frac{x_1}{n} \cdot g_{x_1} \qquad (8.15)$$

where $Q \equiv Dx$ and t_1 is the terminal diffusion date.

The first point to make is that as $n \to \infty$ Equations 8.14 and 8.15 tend to Equations 8.12 and 8.13 with $n = 1$, i.e. the perfect foresight path for a large number of producers approaches the myopia path with a single supplier. The logic of this result is that, as n increases, the supplying industry comes nearer to being competitive. Then all profits from the adoption of the new technology goes to the users, who choose a rate of take-up to maximize total rents – thus emulating the choice of the single monopoly supplier with myopic buyers.

For $2 \leqslant n \leqslant \infty$ on the perfect foresight path, Equations 8.14 and 8.15 indicate that at t_1, $Dc = 0$ and thus $t_1 = \hat{t}$. Thus the terminal data is unaffected by n, and is still $\hat{t}$. From Equation 8.15 a sufficient condition for $x(t)$, $t \leqslant \hat{t}$ to increase with n is that $d[xg(x)]/dx > 0$, and $d^2[xg(x)]/dx^2 < 0$. This condition is analogous to that of positive but declining marginal revenue in a static market and we will assume that it holds. Then a greater number of suppliers implies increased usage for all t due to the lower price trajectory of the more competitive environment and we can state:

Proposition 1: Multiple suppliers and buyers' perfect foresight yield an open loop equilibrium with sales until $\hat{t}$ and higher sales at any point in time the greater number of suppliers. As the number of suppliers becomes infinite the perfect foresight path approaches the path under myopia with one supplier.

Under myopia with multiple suppliers, we have from Equation 8.12 that for all $t < t_1$, $rc - Dc = g(x) - (n - 1) Qg_x/rn$, where as with a single supplier $rc - Dc = g(x)$. Given $g_x < 0$, x will be greater for all $t < t_1$ with a greater number of suppliers. One solution to Equations 8.12 and 8.13 has $Dc = 0$ and $Q = 0$ at t_1, which implies $t_1 = \hat{t}$. Although other solutions to Equations 8.12 and 8.13 which involve $Dc < 0$ and $Q > 0$ at t_1 (and thus $t_1 < \hat{t}$) are possible, we cannot consider any t_1 thus generated as a valid terminal point since such a solution would not constitute an open loop equilibrium. The reason for this is that any one firm would then take other firms' production to end at t_1 and would conjecture additional profits for itself by remaining in production at t_1 and waiting for cost savings to generate further profitable sales. As all firms would have similar conjectures, an open loop equilibrium can only be characterized by production continuing until $\hat{t}$ after which no further cost reductions are possible. We can thus state:

Proposition 2: Multiple suppliers and buyers' myopia yield an open loop equilibrium where accumulated sales are greater at any time before $\hat{t}$ than in the case of a single supplier.

We may thus state that a more competitive market structure in the supply industry encourages greater use of the new technology. However, in Ireland and Stoneman (1986), we also point out that within the confines of the model, the welfare optimal path is the path that would be generated by a monopolist supplier facing myopic buyers. As we have stated, the same path is generated by a supplying industry facing buyers with perfect foresight as $n \to \infty$. As n increases diffusion speeds up and we may thus state that an increase in n is welfare-improving if buyers have perfect foresight but will produce too fast a diffusion if buyers are myopic.

This result and the model, however, take the existence of the technology as predetermined. As we have indicated, the profits of the suppliers will depend on the foresight of buyers and the number of suppliers. As n increases suppliers' profits decrease and profits are also lower under perfect foresight. These profits (or the expectation of them) represent the incentive to develop the technology in the first place. In a further paper with Norman Ireland discussed in Stoneman (1987), we consider this complication, basically arguing that given free entry into R&D for developing new technologies, n will be endogenous, as well as the nature of the technology to be diffused. Moreover, we point out there that the effectiveness of the patent system will be an important factor determining n and thus the number of suppliers of the new technology and the diffusion path.

8.4   Another look at the demand side: feedback from use to profit gains

Thus far we have modeled the demand side as if the gain to any individual adopter is independent of the number of adopters to date. A more Schumpeterian assumption would relate the gain from adoption to the extent of adoption so far. This is the basis of Reinganum's (1981) model. In essence, assume that apart from adoption dates, all adopters are the same. Let a firm using the new technology make profit $\Pi_1(M, N)$, and a firm using old technology make profit $\Pi_0(M, N)$ where M is the number of adopters and N the number of firms in the industry. Reinganum *assumes* that

$$\Delta\Pi(M, N) \equiv \Pi_1(M, N) - \Pi_0(M, N) \tag{8.16}$$

decreases with M. With an adoption rule that a firm adopts if $p_t <$ $\Delta\Pi(M_t, N)$, one may then trace out a diffusion path as p_t falls over time. As would be obvious, if $\Delta\Pi$ is not a monotonic function of time, the model is much more difficult to operate. Moreover Fudenberg and Tirole (1985) have criticized Reinganum for not considering pre-emption and Quirmbach (1986) has also criticized the Reinganum model.

Rather than pursue these criticisms we will, on the altar of simplicity, look them straight in the eye and pass on to consider a simple model that will contain the essence of the argument. Consider an industry with N firms of whom at any time M_t are the users of the new technology. The new technology yields production costs c_1, the old technology cost c_0, $c_0 > c_1$. There is a linear demand curve

$$Z = a - bQ \qquad a, b > 0 \tag{8.17}$$

where $\quad Q = \sum_i q_i \qquad i = 1, \ldots N$

$\qquad Z$ = industry price
$\qquad q_i$ = output of firm i

Ignoring fixed costs, users of both types of technology maximize profits

$$\eta_i = (a - bQ) q_i - c_k q_i \qquad k = 0, 1 \tag{8.18}$$

assuming Cournot conjectures. This yields

$$q_0(M, N - M) = \frac{a - (M + 1)c_0 + mc_1}{b(N + 1)} \tag{8.19}$$

$$q_1(M, N - M) = \frac{a + (N - M) c_0 - (N + 1 - M) c_1}{b(N + 1)} \tag{8.20}$$

as the output of old and new technology firms. We assume that c_1 is not low enough to drive $q_0(M, N)$ negative. Then

$$q_1 - q_0 = \frac{c_0 - c_1}{b} \tag{8.21}$$

and

$$Q = Mq_1 + (N - M)q_0 \tag{8.22}$$

We will assume that adopters are completely myopic and consider that the profit gain at the date of adoption will last forever. With this assumption we may look at the period profit gain without the complication of integrating over time. Then

$$\eta_1 - \eta_0 = (a - bQ) (q_1 - q_0) - c_1 q_1 + c_0 q_0 \tag{8.23}$$

$$= \frac{a}{b}(c_0 - c_1) - \frac{M}{b}(c_0 - c_1)^2 - Nq_0(c_0 - c_1)$$

$$- \frac{c_1(c_0 - c_1)}{b} - c_1 q_0 + c_0 q_0$$

and as can be seen as M increases the profit gain from adoption decreases, for

$$\frac{d(\eta_1 - \eta_0)}{dM} = -\frac{1}{b}(c_0 - c_1)^2 \left(\frac{2}{N+1}\right) < 0 \qquad (8.24)$$

Looking at the effect of market structure on diffusion

$$\frac{d(\eta_1 - \eta_0)}{dN} = -q_0(c_0 - c_1) + \frac{dq_0}{dN}(1 - N)(c_0 - c_1) \qquad (8.25)$$

Given

$$\frac{dq_0}{dN} = -\frac{q_0}{N+1};$$

$$\frac{d(\eta_1 - \eta_0)}{dN} = (c_0 - c_1)(q_0)\left(\frac{-2}{N+1}\right) < 0 \qquad (8.26)$$

Thus in this simple framework the profit gain from adoption decreases as N increases. With this result, for a given cost of acquiring the new technology, M_t will be lower as N is higher, i.e. lower concentration (higher N) generates a lower number of users of the new technology.

This is not, however, the main reason for analyzing this model. I am more interested in using it to explore how market structure of the using industry will vary as diffusion proceeds. Basically if M is zero (all firms use old technology) and all are assumed the same, concentration as measured by CR_k would be k/N. As diffusion proceeds however, q_0 and q_1 change and thus so would measured market structure.

Now we may state that

$$CR_k = \frac{Mq_1 + (k - M)q_0}{Mq_1 + (N - M)q_0} \text{ for } k \geqslant M \qquad (8.27)$$

and

$$CR_k = \frac{kq_1}{Mq_1 + (N - M)q_0} \qquad \text{for } k < M \leqslant N \qquad (8.28)$$

For $k \geqslant M$

$$CR_k = \frac{[M(c_0 - c_1)/b] + kq_0}{[M(c_0 - c_1)/b] + Nq_0} \qquad (8.29)$$

and

$$\frac{dCR_k}{dM} > 0$$

For $k < M$

$$CR_k = \frac{kq_0 + k(c_0 - c_1)/b}{M(c_0 - c_1)/b] + Nq_0} \tag{8.30}$$

and

$$\text{sign } \frac{dCR_k}{dM} = \text{sign } \left\{ \frac{(c_0 - c_1)^2}{b^2} \left[\frac{k(-1 - M)}{N + 1} \right] - \frac{kq_0(c_0 - c_1)}{b} \right\}$$
$$= \text{negative}$$

Thus for $k \geq M$, CR_k increases with M, for $k < M$, CR_k decreases with M. The actual signs themselves are not important. The critical issue is that measured concentration changes as diffusion proceeds. Market structure results from the diffusion process, but more importantly, an observation on market structure at a moment in time is just one point on a path of market structure, and this one point may not be a true guide to the structure of that industry.

Thus far I have concentrated on market structure. What then of the relationship of adoption dates to firm size? In this model as M increases both q_0 and q_1 fall. It is thus clear that later adopters will have lower values of q_0 than earlier adopters, and thus smaller firms are associated with later adoption. However, as should be clear, the relationship is not one in which small size causes late adoption, but one in which late adopters are small because they are late adopters.

8.5   Concluding remarks

We have explored the relation between diffusion, market structure and firm size in several modeling frameworks. There is no consistent set of findings from these models, but it is too early in this analysis to expect consistency. Perhaps of more interest here is to conclude by considering two other issues – intra-firm diffusion and product differentiation, this latter being an essential part of any study of the diffusion of product innovations.

The above discussion has concentrated on diffusion across firms and assumed that firms only ever buy one unit of the new technology. Relaxing this assumption raises questions on how market structure and firm size will be related to the rate at which new technology is adopted within the firm. This is not an issue that appears to have been much discussed in the literature, but should be amenable to analysis.

In a recent paper (Stoneman, 1989) I have looked at diffusion in a model of horizontal product differentiation. In that paper I suggest that if brands are controlled by different oligopolists rather than a single brand monopolist then the diffusion will be affected by this. In particular I suggest that there will be an end game under oligopoly, and diffusion under oligopoly will be more extensive than under monopoly. For most of the diffusion path however, the differences in market structure will have no effect. These are, however, conjectures, and represent an avenue for future research.

References

David, P. (1969), *A Contribution to the Theory of Diffusion*, Stanford Centre for Research in Economic Growth, Memorandum No. 71, Palo Alto: Stanford University.

Davies, S. (1979), *The Diffusion of Process Innovations*, Cambridge: Cambridge University Press.

Fudenberg, D. and J. Tirole (1985), "Preemption and rent equalization in the adoption of new technology", *Review of Economic Studies*, vol. 52, pp. 383–401.

Ireland, N. and P. Stoneman (1986), "Technological diffusion expectations and welfare", *Oxford Economic Papers*, vol. 38, pp. 283–304.

Quirmbach, H. (1986), "The diffusion of new technology and the market for an innovation", *Rand Journal of Economics*, Spring, vol. 17, pp. 33–47.

Reinganum, J. (1981), "Market structure and the diffusion of new technology", *Bell Journal of Economics*, vol. 12, pp. 618–24.

Romeo, A. (1977), "The rate of imitation of a capital embodied process innovation", *Economica*, vol. 44, pp. 63–9.

Stoneman, P. (1987), *The Economic Analysis of Technology Policy*, Oxford: Oxford University Press.

Stoneman, P. (1989), *Technological Diffusion, Horizontal Product Differentiation and Adaptation Costs*, mimeo, February, Coventry: University of Warwick.

9

Research diversity induced by rivalry

John T. Scott*

9.1   Overview

The diversity synonymous with the rivalry in Schumpeter's (1942) creative gale of competition is the subject of this chapter. Scholars have often asked whether competition or Schumpeterian monopoly is more likely to promote technological progress. Arguably Schumpeter's vision, though, is that, in an evolutionary context, competition and monopoly are part of the same process. Greater competition implies a greater incentive to use divergent research strategies to lessen the anticipated erosion of rents caused by competition among rival innovations in the post-innovation market. Thus, the rivalry of numerous competitors causes them to establish unique research strategies defined by differences in the generic innovation sought and differences in the method of search. Greater structural competition,

* The earlier version of this chapter was prepared for the Conference on Innovation and Technological Change, Wissenschaftszentrum Berlin für Sozialforschung, August 10–11, 1989. This version sharpens the explanation of firms' decisions to "locate" differently their R&D strategies, and it also adds some new evidence showing a correlation between structural competition and the significance of monopoly-like distinctions among sellers. I thank William L. Baldwin, Paul A. Geroski, Steven Klepper and Geoffrey Woglom for comments, and I am especially indebted to F. M. Scherer for helpful correspondence and for granting me access to the research file that he developed about the systems orientation of patent portfolios. Joseph Cholka and George Pascoe carried out the data processing using the Line of Business (LB) data from the United States Federal Trade Commission (FTC). The FTC's Disclosure Avoidance Officer has certified that these results do not identify individual company LB data. The conclusions presented herein are those of the author and have not been adopted by the FTC or any entity within the Commission.

then, implies a larger number of sellers, monopoly-like in the sense that each has a unique R&D strategy.

Section 9.2 describes industrial research and development (R&D) in a way allowing understanding of research diversity. Section 9.3 explains why rivalry increases such diversity. Section 9.4 offers evidence of firm effects in R&D strategies and a correlation between structural competition and the significance of monopoly-like distinctions among sellers. Section 9.5 concludes, noting that, analogous to Chamberlin's (1933) observations about competition and product differentiation, the diversity created by R&D competitors can be socially beneficial.

9.2   Research discovers component gestalt

Innovations are new bundles of components that work together; the components have consistent attributes. The integration of an innovation's components achieves *component gestalt* – the necessary integration of components. Basic research and the creative inventive act conceive the components in their essential working configuration. Development refines their integration. With the development process, enough is known so that trial and error, although typically costly, focus on the possibilities for performance attributes of the R&D project's components.[1]

A component here could be a part of what might itself be viewed as a component. For example, the development of a hypersonic transatmospheric aircraft would entail development work on the aircraft's exterior. That work could be construed as a project in itself requiring development of "components" including the material, the fabrication process and the design.

A complete R&D product requires research on all components of the project and success requires an outcome for which the components' performance attributes are consistent. For example, the research on fabrication must discover a technique compatible with the material developed. A technique requires a material of particular weight and strength. Further, the configuration of the aircraft must be consistent with certain types of materials and fabrication techniques, because the design requires materials and a fabrication process with particular attributes.

The extent of component gestalt achieved can be high or low. Table 9.1 illustrates a project that has five components. The component gestalt is greatest when all five components mesh; it is lowest when none of the components function together. Letting x denote the number of distinct project components – distinct in the sense that they do not mesh – x

Table 9.1   The extent of component gestalt $(C = 5; Z = 840)$.

x^a	General patterns[b]	Occurrences[c] $(= y)$
1	5	840
2	4, 1	3 523 800
	3, 2	7 047 600
3	3, 1, 1	$5.905\,888\,8 \times 10^9$
	2, 2, 1	$8.858\,833\,2 \times 10^9$
4	2, 1, 1, 1	$4.943\,228\,9 \times 10^{12}$
5	1, 1, 1, 1, 1	$4.132\,539\,4 \times 10^{14}$

$$\Sigma y = 4.182\,119\,4 \times 10^{14} = Z^C = (840)^5$$

Notes: [a] x denotes the number of components that are distinct in the sense that they do not mesh.
[b] The general patterns for sets of distinct components consistent with x are shown as integers ordered from largest to smallest with each integer being the number of consistent components with a particular set of performance attributes.
[c] The number of ways that each particular general pattern could occur is shown in this column.

ranges from 1 through 5 as the project's component gestalt varies from high to low. When x is 1, all five components work together. If x were 2, either four components function together and one does not mesh, or one group of three compatible components and another group of two constitute the outcome. Thus, when $x = 2$, the general patterns for sets of distinct components are 4 and 1 or 3 and 2. When $x = 5$, the only general pattern is 1, 1, 1, 1 and 1. Each of the five components is distinct – does not mesh. To mesh, their performance attributes must be compatible. If the project is the development of a pollution-free engine, and the fuel developed burns hot, but the combustion chamber cannot take the heat given the design and the materials, then the fuel and the combustion chamber do not mesh.

Let C denote the number of project components, where the project could be the grand project (such as the plane) or a proper subset (such as the fuselage). Let Z be the number of ways each component could turn out for the set of characteristics that must mesh across all components. That meshing or consistency is necessary for a successful innovation. Yet for the typical R&D project given C and Z only a small proportion of the Z^C possible outcomes will exhibit the necessary consistency.

Let y be the number of ways that each particular general pattern for sets of distinct components could occur. Let the set of integers, ordered from largest to smallest, denoting any such general pattern be

$$q = \{q_{11}, q_{12}, \ldots, q_{1Q1}, q_{21}, q_{22}, \ldots, q_{2Q2}, \ldots, q_{v1}, q_{v2}, \ldots, q_{vQv}\},$$

where there are $Q1$ instances of the largest integer, $Q2$ of the second, and so on until Qv instances of the smallest integer. The $\Sigma_{ij}q_{ij}$ is equal to C, the number of project components associated with the R&D project. And, for any general pattern, there are v distinct integers and $x = \sum_{h=1}^{v} Qh$ distinct project components. Then, with $\Omega_{s,t}$ denoting the combination of s things taken t at a time, the number of occurrences y for any given general pattern will, since $q_{ij} = q_{ik} = q_i$, be:

$$y = \prod_{n=1}^{v} \left[\Omega_{Z - \underset{(i<n)}{\Sigma} Q_i, Q_n} \cdot \Omega_{C - \underset{(i<n)}{\Sigma} (Q_i \cdot q_i), Q_n \cdot q_n} \cdot \prod_{i=1}^{Qn} \Omega_{q_n \cdot i, q_n} \right] \qquad (9.1)$$

For a non-trivial R&D problem, Z is large relative to C, and consequently, most of the possible development outcomes have low degrees of component gestalt.[2] The absolute number of development outcomes with a high degree of component gestalt is typically large and increases with complexity (measured directly by Z). As a proportion of the total number of outcomes, however, the high component gestalt cases are rare – increasingly rare as project complexity increases. The small proportion of the sample space associated with high component gestalt shrinks as the complexity of the development task increases. Table 9.2 illustrates how complexity increases the large proportion of the sample space for which none of the components mesh.

The foregoing abstraction, which is used throughout this essay, is consistent with accepted observations about innovation:

1. Genius, or at least the flash of insight, is necessary to conceive the C components of an innovation; accumulated knowledge and trial and error are necessary to develop their consistency.

Table 9.2 For an R&D project with five ($C = 5$) project components, the proportion of the sample space for which there is no component gestalt – i.e. all project component vectors are distinct – as the project becomes more complex (as Z increases).[a]

Z	y/Z^C
100	0.903 45
1 000	0.990 035
10 000	0.999
100 000	0.999 9
1 000 000	0.999 99

Note: [a]For comparison, in this $(1, 1, 1, 1, 1)$ case where $x = 5$ and $Z = 840$ in the numerical example provided in Table 9.1, $y = 4.132 539 4 \times 10^{14}$ and $y/Z^5 = 0.988 145$.

2. Unless basic science and applied experience are strong, the development process for complex innovations entails costly trial and error to uncover consistent outcomes, because the proportion of the sample space for which components are consistent is small.
3. There are an immense absolute number of potential solutions to an R&D problem; hence, there is room for diversity of solutions.

9.3 Rivalry and diversity

A research strategy includes a choice of the number of components C and of complexity Z. My hypothesis is that competing firms, with identical capabilities and opportunities, choose different R&D strategies to increase the expected value of their innovative investment. Given that patents, trade secrets or first-mover advantages protect innovations from imitation, and given that innovating firms would not anticipate a sole winner of a patent on the innovation sought (but rather patented, competing substitutes), the firms would want research strategies that make more likely a single winner, rather than multiple winners, for each R&D rivalry. Greater competition implies greater incentive to use such divergent research strategies, because in their absence the erosion of rents, caused by competition among rival innovations in the post-innovation market, would be more severe.

A simple model of independent R&D trials explains why R&D rivals would have an incentive to create, for any particular innovation, a competition that only one firm will win. Research entails an outlay in anticipation of an uncertain return. Suppose that a unit of R&D input allows one trial with the probability of success equal to $n_1/(n_1 + n_2)$, where a trial has n_1 chances for success and n_2 chances for failure. Since not all consistent outcomes have value, $n_1 < Z$, the number of consistent outcomes.

Let V denote a monopolist's total private value of successful innovation, regardless of the number of successful, consistent component configurations introduced. Thus, whether one, two or more of the n_1 potential successful innovations were introduced, multiple innovations would share the market and total private value would be V.

What then is the monopolist's expected net benefit, B, from innovative investment? It is:

$$B_m = \left\{ \sum_{i=1}^{f} (V)\Omega_{f,i}[(n_1^i n_2^{f-i})/(n_1 + n_2)^f] \right\} - \text{Cost}(f) \qquad (9.2)$$

where f is the number of research trials and $\Omega_{f,i}$ denotes the number of combinations of f things taken i at a time. The formulation follows because $(n_1^i n_2^{f-i})/(n_1 + n_2)^f$ is the probability of each set of i successful trials and $\Omega_{f,i}$ gives the number of such sets. The monopolist conducts trials up to the point where one more would add less to expected benefit than to cost. Since, as trials increase, the marginal benefit of a trial is diminishing, there is an optimal number of trials (Nelson, 1961; Nelson, 1982; Evenson and Kislev, 1976).

We can now depict a non-cooperative equilibrium that is identical to the monopoly solution. If each firm conducts only one trial, then with $V(i)$ denoting the value of innovative investment to an individual firm when i firms succeed in discovering one of the n_1 substitutes, expected profit for the individual firm given f firms in the market is:

$$B_c = \left\{ \sum_{i=1}^{f} V(i)\Omega_{f-1,i-1}[(n_1^i n_2^{f-i})/(n_1 + n_2)^f] \right\} - \text{Cost}(1) \quad \textbf{(9.3)}$$

since, as before, the expression containing n_1 and n_2 denotes the probability of each set of i successful trials and since $\Omega_{f-1,i-1}$ is the number of ways the individual firm can be among i winners.

Now, we have our result: with each firm maximizing its expected profit under the assumptions that it will do $(1/j)$th of the total innovative investment (where j is the number of firms undertaking innovative investment) and that its post-innovation market share will be $1/i$ (where i is the number of firms innovating in a given state of nature), the total amount of innovative investment in the industry will be *in equilibrium* at an amount equal to the amount undertaken by a monopoly if the number of firms equals the number of units of innovative investment that would be undertaken by the monopoly.

For the proof of our result: first, $B_m(f) = f[B_c(f)]$, since each term within the summation sign for B_c is $(1/f)$ times the corresponding term for B_m. From Chamberlin (1929) and Osborne (1976), we know (given our assumptions) that $V(i) = V/i$. Then, for B_c the terms are:

$$V(i)\Omega_{f-1,i-1}[(n_1^i n_2^{f-i})/(n_1 + n_2)^f] \quad \textbf{(9.4)}$$

$$= \frac{V}{i} \frac{(f-1)(f-2)\ldots[(f-1)-(i-1)+1]}{(i-1)(i-2)\ldots(1)} \times$$

$$[(n_1^i n_2^{f-i})/(n_1 + n_2)^f] = (1/f)V\Omega_{f,i}[(n_1^i n_2^{f-i})/(n_1 + n_2)^f]$$

Since by assumption $\text{Cost}(f) = f(\text{Cost}(1))$, we then have $(1/f)B_m(f) = B_c(f)$. But then if the number of firms is equal to the f that maximizes B_m, those f firms will find that they have no incentive to change their innovative investment. To undertake more investment *symmetrically* clearly lowers profits $B_m(f) = fB_c(f)$ and, in fact, individual firm profits, since $2fB_c(f+f) < fB_c(f)$ implies $2B_c(f+f) < B_c(f)$. To undertake less investment would mean abandoning a profitable project. Thus, this non-collusive outcome, identical to the monopoly outcome, is an equilibrium and the proof of our result is complete.

The exposition of the result makes clear why rivalry in research will usually not result in an equilibrium identical to the monopoly solution. Our conditions included

(a) symmetric innovative investments for each of the firms, with their number exactly equal to the ratio of the monopolist's number of trials to the number of trials undertaken by each competitor,
(b) no scale economies for multiple trials, and
(c) Chamberlin's (1929) or Osborne's (1976) non-cooperative joint profit maximizing equilibrium in the post-innovation market.

Typically we would expect these conditions to be violated, and we would expect that rivalry in R&D would cause firms to expect rivalry among competing substitutes in the post-innovation market.

Thus, with competition, the total expected profit would typically be less than what the monopolist expects. Of course, many theorists have shown that, in spite of the low profits, the competitors in such a game might well spend more on R&D than would the monopolist (Baldwin and Scott, 1987). But this is an unattractive game from a private standpoint, and consequently, R&D competitors have an incentive to incur costs that a monopolist would not want to incur, solely to change the R&D game for a particular innovation into one that only a single firm could win. For then, given that the rivals do all pursue a unique form of the innovation and do not simply allow each other to pursue particular innovations alone, each firm's innovation investment costs would yield the expected benefit $(1/j)$th of the monopolist's benefit given j trials, an expectation greater than the expected benefits with competition:

$$\left\{ \sum_{i=1}^{i} V(i)\Omega_{j-1,i-1} \left[(n_1^i n_2^{j-i})/(n_1 + n_2)^j \right] \right\} \tag{9.5}$$

given that $V(i) < V/i$.

To create a research game that only one firm could win, the rivals would pursue diverse strategies aimed at producing a unique product unlikely to be considered a mere substitute for competing innovations, but instead likely to have a decisive advantage that would drive other innovations from the post-innovation market. Even if one R&D strategy seems more promising than the next best alternative, once the preferred strategy has been pre-emptively taken by a rival, a firm can prefer an alternative that has some chance of finding a dominant innovation. However, our discussion of the sample space suggests the immense number of possible solutions may yield numerous strategies that are *a priori* equally likely to succeed. A monopolist would not have an incentive to incur the costs of ensuring diverse outcomes for trials in order to increase the likelihood of just one dominant product, since regardless of the number of trials producing successful, substitutable innovations, the monopolist gains the same expected benefit. Substitutable products commercialized as a result of the monopolist's R&D trials will not create rent-eroding price competition in the post-innovation market.

9.4   Evidence of competition-induced diversity

The evidence below shows that rivalry and diversity are synonymous; however, the effect is not captured by traditional correlations of R&D behavior and market structure. Instead, given competitors in an area of R&D, we observe distinct R&D strategies, including differences in systems orientation, R&D intensity and purposive diversification of R&D.

If rivalry does induce diversity, we should observe differences in R&D strategies of firms in the same area of R&D. Our test assumes that, in the absence of random error, a complete model would exhibit a definite relationship among n non-stochastic variables, $(y_1, y_2, \ldots, y_i, \ldots, y_n)$, describing industries, firms and lines of business, where a line of business (LB) is the operation of a firm in a particular industry category. That relationship would be $\Gamma(y_1, y_2, \ldots, y_i, \ldots, y_n) = 0$, where for example one of the variables is the LB variable y_i which is the proportion of the LB's patent portfolio devoted to systems or subsystems, i.e. developments integrating components (in the ordinary sense). Then, we observe y_i where $y_i = \Psi(y_i, y_2, \ldots, y_{i-1}, y_{i+1}, \ldots, y_n) + \varepsilon$, where ε is homoskedastic random error, with mean zero, uncorrelated with y_j, $j \neq i$, and where the function Ψ is approximated well by a function linear in its parameters but not in its variables. Then estimations ask whether LB, firm and industry variables explain, in ways predicted by theory, a substantial portion of the variance in y_i.

Schumpeterian hypotheses about LB size, firm size, diversification and industry concentration have not explained much of the variance in R&D intensity.[3] Our discussion, however, suggests new tests using y_i. First, it predicts strong firm effects since firms in the same industries will choose different strategies (including different Cs and Zs) and thus have different proportions of their patent portfolios in systems. The various strategies should vary markedly in outlays for R&D; they span different sets of industry categories.

Systems orientation of patent portfolios

Conceivably, the often examined characteristics of firms could affect their choice of systems orientation. Very large firms and diversified firms might be more likely to undertake the more costly development projects that attempt to integrate greater numbers of components. Smaller and less diversified firms might tend to focus on the less costly, less integrated projects as well as on the less costly, more basic research. That would be consistent with, but not the same as, the observation of Scherer (1980, pp. 416–17) and Mansfield *et al.* (1977a) that smaller firms may tend to *invent* but larger firms tend to *develop* industrial products and processes. Whether the choice of C and Z (the number of components and the complexity of their integration) in *development* projects differs significantly across firms of different size is the issue here. Since the proportion of the sample space for which component gestalt is high is typically quite small for complex systems developments, greater R&D outlays will be necessary to find solutions not among the relatively plentiful low gestalt solutions.

The theory also predicts strong industry effects. Industries differ in their value and cost functions for degrees of component gestalt, and that will cause differences in systems orientation, apart from any effects of differences in competition.

While we can extend earlier studies by asking if the Schumpeterian characteristics of size, diversification and concentration are correlated with systems orientation, the new approach is to ask if more rivalry, i.e. more firms doing R&D, does lead to diversity in the form of a filling of more sample space niches – indicated if we discover firm effects in R&D activity. Rivalry can lead not only to more socially desirable levels of R&D, but to more socially desirable composition – indeed to less duplication in the sense that less of the R&D effort is focused on a given strategy or portion of the sample space. Perhaps seller concentration is correlated with rivalry and is therefore important for understanding differences in systems R&D across industries. But the theory predicts intra-industry differences among firms rather than a conventional partial correlation between firm behavior and seller concentration.

F. M. Scherer provided patent characteristics data for the mid-1970s patents of 1,819 Federal Trade Commission (FTC) lines of business (LBs) described and studied in Scherer (1983). His data are used here to explore the extent to which LB, firm and industry characteristics are correlated with systems developments among large US manufacturing firms.The LB level variables include a measure of y_i, the dependent variable, denoted *SYST*; it is the fraction of an LB's patents pertaining to systems or subsystems. *SYST* is measured as a proportion with mean equal to 0.39. All other variables are described in detail in the Appendix. They include two additional LB variables. One is the ratio of contract R&D outlays to total R&D outlays; the other is the sales in an LB. Company level variables are a measure of diversification and a measure of firm size. Additionally, D_i denotes a dummy variable for the ith firm. A measure of seller concentration is an industry level variable, and T_j denotes a dummy variable for the jth industry.

Scherer (1983, pp. 124–5) uses the 1,819 observation sample and regresses *SYST* on a measure of LB sales and technological class dummies. After duplicating those results, I extended the analysis as follows.[4]

Equation 9.6 shows that industry effects alone explain 49 per cent of the variance in *SYST*, and Equation 9.7 shows that firm effects alone can explain 46 per cent of *SYST*'s variance.

$$SYST = z + \sum_{j=2}^{238} \gamma_j T_j \qquad (F = 6.5)^a \qquad (9.6)$$

F-value = 6.5^a,R^2 = 0.49, degrees of freedom = 1,581

$$SYST = z + \sum_{i=2}^{397} \beta_i D_i \qquad (F = 3.1)^a \qquad (9.7)$$

F-value = 3.1^a, R^2 = 0.46, degrees of freedom = 1,422

In Equation 9.8, we see that firm and industry effects together explain 64 per cent of the variance in *SYST*.[5]

$$SYST = z + \sum_{i=2}^{397} \beta_i D_i + \sum_{j=2}^{235} \gamma_i T_i \qquad (9.8)$$
$$(F = 1.3)^a \qquad (F = 2.6)^a$$

F-value = 3.4^a, R^2 = 0.64, degrees of freedom = 1,188

When interpreting the results, remember that when firm (industry) dummies are included, firm (industry) level variables cannot be. The effects of the latter are nested within the effects controlled by the dummies.[6] Further, remember that I am describing correlations and their strengths, not identifying the direction of causal links. Formally, the direction of causality is not specified, but I assume that the error in an equation is uncorrelated with the included variables.

Unlike models of symmetric Nash equilibria, our description of industrial R&D implies that firm effects as well as industry effects will be important. Further, our hypotheses about "Schumpeterian" variables imply both firm and industry effects will be present and important in *SYST*. They clearly are. Beyond that observation, we cannot proceed with confidence to unambiguous *ceteris paribus* effects of particular firm or industry variables, because as shown in the Appendix, the traditional variables explored here explain only a small portion of the variance explained by the dummy variables, and hence, by all relevant firm and industry variables. The results, as shown in the equations provided in the Appendix, are as follows.

Although left-out industry variables no doubt affect the coefficient, even after control for firm effects, seller concentration is associated with relatively extensive systems developments. Possibly the correlation reflects fewer niches to be filled in industry categories where systems development is important, but the left-out variable problem renders any story inconclusive. Here, as with the variables studied subsequently, if we had found that the particular industry (firm) level variable or variables explored explained a large portion of the variance explained by the industry (firm) dummies, we could have had some confidence in the correlations as reflections of *ceteris paribus* effects. However, seller concentration alone explains at most only 1.6 per cent of the variance in systems orientation.

More diversified firms did more systems work, but that effect is because they also tend to be in industries where more systems work is done, and diversification explains less than 1 per cent of the variance in *SYST*. An increase in overall firm size is associated with a lower proportion of an LB's patents devoted to systems, but the effect is small, barely significant, and explains far less than 1 per cent of the variance. As we would expect, contract work tends to support systems developments (since federal government contracts often support such work), but firm and industry effects explain the presence of contracted R&D, which when entered alone explains about 3 per cent of the variance in *SYST*. LB sales is not significant in complete specifications, adding virtually nothing to explanatory power.[7]

R&D intensity

Evidence of firm effects in R&D intensity also supports the idea that firms in the same industry categories pursue different R&D strategies. The importance of firm versus industry effects has been documented in Scott (1984). Table 9.3 shows new evidence. Instead of controlling for FTC four-digit industry effects as in Scott (1984), the effects of the differences among 20 SIC two-digit manufacturing industries and the 127 groups are controlled, each group being a related set of FTC four-

Table 9.3 The dependent variable is *RS* (LB total R&D)/(LB sales) for the 2,315 observations on LBs existing for all three years and in which some R&D was performed. These observations are for the 352 firms in the sample that reported R&D throughout the sample period for one or more FTC four-digit manufacturing categories.[a]

	(1)	(2)	(3)	(4)
351 firm effects	$F = 1.7$			fitted last $F = 1.5$
19 two-digit industry effects		$F = 14$		fitted last $F = 4.7$
127 group effects			$F = 2.3$	fitted last $F = 1.7$
Intercept	z	z	z	z
F-ratio for the equation	1.7	14	2.3	2.0
R-square	0.23	0.10	0.12	0.36
Degrees of freedom	1 963	2 295	2 187	1 817

Note: [a]All *F*-ratios are highly significant, far beyond the conventional 0.01 level. To reduce the size of the $X'X$ matrix for computational purposes, in some specifications firm effects were absorbed, and then the R^2 and *F*-tests were constructed using the information from several regressions. To get Column 4, with variables in deviation form, we regressed *RS* on the industry dummies, *RS* on the group dummies, *RS* on the industry dummies and the group dummies. And then, with the variables not as deviations, we regressed *RS* on the industry dummies and the group dummies. In other words, we fit everything except the group effects, everything except the industry effects, everything, and then everything except the firm effects. We then used the sums of squares thereby obtained with the appropriate degrees of freedom to construct Column 4. For a complete description, in a simpler context, of how these regressions are used to get the *F*s for effects fitted last, see Scott (1984, p. 244).

digit industries found in Scott and Pascoe (1987). Each group consists of a set of industry categories for which R&D activities were deemed complementary, with shared R&D facilities or spillovers (among the categories) of knowledge generated by R&D. The dependent variable is an observation of R&D intensity for 1974–76 for each line of business (LB), i.e. the activities of a firm in an FTC four-digit manufacturing category, in the Scott and Pascoe (1986) sample for which there was R&D activity throughout the sample period. R&D intensity is measured as the ratio of LB total (applied) R&D to LB sales.

Table 9.3 is consistent with the prediction that firms differ in their R&D intensity, even after industry or group effects are controlled. In Section 9.3, we predicted that firms will differ simply because they choose to differ, thereby increasing the likelihood that for any given project only one competitor will win the R&D contest. In Table 9.3, the firm as well as the industry and group effects are significant.[8] Yet previous studies of R&D intensity have shown that the traditional variables identified as factors in Schumpeterian competition explain very little of that systematic variance in R&D intensity.[9]

Our hypothesis suggests as a conjecture that firm effects in R&D intensity will be more pronounced in markets with more competitive structures (in the sense of more firms), because there is more need for a firm to strive for a different outcome to avoid erosion of rents in the post-innovation market. Thus, somewhat paradoxically, a more competitive structure would imply more monopoly in the sense of more significant monopoly-like differences among sellers. We would find, then, a positive correlation between insignificance of firm effects and the magnitude of seller concentration. To test this conjecture, let *INSIG* measure directly the insignificance of the firm effects in R&D intensity within each of the twenty two-digit SIC industries. *INSIG* is the probability of a larger *F*-value for the *F*-test against the null hypothesis of no firm effects within each two-digit SIC industry. *INSIG* is taken from Scott (1984, Table 10.4, pp. 238–40). Further, let *ACR* be the adjusted concentration ratio in percentage terms for each of the twenty SIC two-digit industries. *ACR* is a sales-weighted average of the underlying four-digit FTC adjusted concentration ratios described in Scott and Pascoe (1986).

Then, the conjecture is that *INSIG* and *ACR* are positively correlated, and we do find that correlation. For the twenty two-digit industry observations, *ACR* explains 20 per cent of the variance in *INSIG*, with a significant positive correlation coefficient of 0.45. Inspection of the plot of *INSIG* on *ACR* suggests that *INSIG* increases at an increasing rate as *ACR* increases. Fitting a quadratic relation gives an insignificant coefficient on the unsquared term; the more parsimonious functional form with only the squared term yields:

$$INSIG = \underset{(t = 0.520)}{0.0629} + \underset{(t = 2.33)}{0.000126\,ACR^2} \tag{9.9}$$

with $R^2 = 0.232$. *ACR*, then, as it ranges from 15.66 to 73.84 in the sample, has a large predicted inverse impact on the significance of differences in R&D intensity. Moving from the least concentrated to the most concentrated two-digit industry increases from 0.0938 to 0.750 the predicted probability of a greater *F*-value for firm effects. Thus, greater competitive structure is associated, in a sense, with more monopoly.

Purposive diversification

Scott and Pascoe (1987) provide another test of the idea that firms in the same industry will pursue different R&D strategies. Evidence presented there shows that firms in a given industry category typically have quite different patterns of purposive R&D diversification and that, even in the same industry category, R&D intensity differs for firms with different diversification patterns. That a firm's idiosyncratic research strategy encompasses multiple industries is consistent with firm effects in the statistical sense: a firm effect is associated with the firm regardless of the industry category where the firm is observed.

9.5   Conclusion

Models of R&D rivalry have typically described symmetric innovative investment outcomes for the firms in an industry and have not addressed the fact of firm effects. The paradoxical fact is that different firms in the same industries pursue different R&D strategies. There are two possible explanations:

(a)   firms are fundamentally different in their capabilities;
(b)   the R&D environment would lead to asymmetrical strategies even if all firms were identical.

The second explanation for firm effects was developed in Section 9.3, but of course both explanations are consistent with evidence of firm effects in Section 9.4. The absence of any identifiable variables that explain the variance in the systems orientation of LB patent portfolios, however, supports the second explanation, because the choice of different strategies may be made by firms otherwise identical, not differing in any LB, firm or industry variables other than those like *SYST* that reflect the deliberate choice of a different strategy in order to avoid rent-destroying competition. Further, the previous studies of R&D intensity have shown that the traditional Schumpeterian variables explain very little of the systematic variance in R&D intensity. Yet, as we have seen, unidentified firm and industry effects do explain a large part of that variance; hence, our examination of R&D intensity also supports the second explanation. Indeed, that explanation – *pure strategic differences among firms with identical capabilities and opportunities* – could also explain the failure of traditional industrial organization models to capture much of the systematic variance in LB profitability (Scott and Pascoe, 1986).

Rivalry among firms, then, causes them to pursue different R&D strategies, because they have an incentive for each innovation project to create a game that only one firm can win. Such research diversity induced by rivalry can improve social economic welfare if a monopolist of R&D would under-invest in diversity. Arguably a monopolist would, from a social standpoint, under-invest in research diversity because much of its social value would not be appropriated by the monopolist.[10] As with innovative investment more generally, rivalry can induce behavior that is socially optimal even though it reduces the total industry profits appropriated privately (Barzel, 1968; Scherer, 1980, p. 431, Note 79; Scott, 1988). Paradoxically, such socially optimal behavior induced by Schumpeterian rivalry occurs in markets that are competitive, in the sense that there are several firms vying for the innovation, yet are monopolistic in the sense that the firms use distinctive R&D strategies and anticipate only one dominant innovation.

Appendix

In addition to *SYST*, the LB level variables include *FED*, the ratio of contract R&D outlays to total R&D outlays, measured as a percentage, and its mean is 6.57. *SALES*, the sales in an LB, is the final LB variable, measured in millions of 1974 dollars, and its mean is 266.38. There are two company level variables. The first is a measure of diversification

$$DIVERS = 1 \bigg/ \sum_{i=1}^{n} s_i^2$$

where s_i is the share of the company's sales in the ith cf its n lines of business, and the second is a measure of firm size *FIRMSIZE* which is total sales for the entire company. For a firm having equal sales in each category, *DIVERS* therefore would show the number of different industry categories in which the firm operated. For the 1,819 observation sample, the mean of *DIVERS* is 6.035. *FIRMSIZE* was measured in thousands of 1974 dollars. Additionally, D_i denotes a dummy variable for the ith firm. *CR*4 is an industry level variable which is a weighted average index of 1972 four-firm seller concentration in meaningfully defined components of the FTC line of business categories. The variable is developed, discussed and used in Scherer (1983, p. 118). It is measured as a percentage, and its mean in the 1,819 observation sample is 43.37. Finally, T_j denotes a dummy variable for the jth industry.

Here, in the order in which they are discussed in the text, are the additional equations estimated. Although to estimate the firm effects in the following specifications the effects were absorbed to reduce the size of the $X'X$ matrix, the F-values for firm effects (as well as the industry effects) are for the effects when fitted last and are derived using additional estimations in the same manner as in Scott (1984, p. 244). Significance levels for a one-tailed test are noted as $a = 0.0001$ or better, $b = 0.05$ or better, $c = 0.10$ or better and a^* is recorded for the F-value that I computed as in Scott (1984, p. 244) and for which I did not have a computer generated probability value. Computing the relevant integral is burdensome given the large numbers of degrees of freedom, but the result is obviously highly significant.

$$SYST = z + \sum_{i=2}^{397} \beta_i D_i + 0.00074\ CR4 \qquad \textbf{(9.A1)}$$
$$(F = 3.0)^a \qquad (t = 1.5)^c$$

F-value = 3.1^a, R^2 = 0.47, degrees of freedom = 1,421

$$SYST = 0.27 + 0.0027\ CR4 \qquad \textbf{(9.A2)}$$
$$(t = 12)^a \qquad (t = 5.4)^a$$

F-value = 29^a, R^2 = 0.016, degrees of freedom = 1,817

As noted above, *FIRMSIZE* was measured in thousands of 1974 dollars. Since when I began this project I was no longer a consultant at the FTC's LB Program and had to write the research design before the program lost its last research assistant, submit it for the research assistant to carry out, and then, because of strict confidentiality requirements, wait to see the results until after they had been cleared through the FTC and after the last research assistant had left the program, I did not realize until receiving the cleared printouts that with that scaling of *FIRMSIZE* and the number of spaces allocated on the printout for a coefficient, the statistically significant (barely – at the 0.10 level for a one-tailed test) negative coefficient for *FIRMSIZE* was reported to me as a negative sign followed by a zero, a decimal point and then a string of seven zeros. I had asked for *FIRMSIZE* as Σ *SALES* which would have scaled it in millions, but the proper variable was taken directly from the reporting form in thousands. I report the coefficient below in Equations 9.A3 and 9.A4 as *d*. The computer did report the ratio of the coefficient to its standard error. Thus, all I can report for this variable is that its effect is negative, barely significant and not large. Its absolute value must be less than 0.0000001. Thus, an increase in overall firm size is associated with a lower proportion of an LB's patents devoted to systems, but an increase in firm size of one billion 1974 dollars, lowers that proportion by less than 0.1. How much less, however, I cannot say.

$$SYST = z + \sum_{j=2}^{238} \gamma_j T_j - 0.00023\ DIVERS \qquad \textbf{(9.A3)}$$
$$(F = 6.4)^a \qquad (t = -0.12)$$
$$- d\ FIRMSIZE$$
$$(t = -1.4)^c$$

F-value = 6.4^a, R^2 = 0.49, degrees of freedom = 1,579

$$SYST = \quad 0.34 \quad + 0.0082 \; DIVERS - d \; FIRMSIZE \quad \textbf{(9.A4)}$$
$$(t = 20)^a \qquad (t = 3.8)^a \qquad (t = -0.33)$$

F-value $= 7.1^b$, $R^2 = 0.0077$, degrees of freedom $= 1,816$

$$SYST = z + \sum_{i=2}^{397} \beta_i D_i \; + \; \sum_{j=2}^{235} \gamma_j T_j \qquad\qquad \textbf{(9.A5)}$$
$$(F = 1.3)^{a*} \quad (F = 2.54)^a$$
$$+ \; 0.00043 \; FED - 0.0000011 \; SALES$$
$$(t = 0.73) \qquad (t = -0.09)$$

F-value $= 3.4^a$, $R^2 = 0.64$, degrees of freedom $= 1,186$

$$SYST = \quad 0.37 \quad + 0.0036 \; FED - 0.0000081 \; SALES \quad \textbf{(9.A6)}$$
$$(t = 37)^a \qquad (t = 7.7)^a \qquad\qquad (t = -0.82)$$

F-value $= 30^a$, $R^2 = 0.032$, degrees of freedom $= 1,816$

$$SYST = z + \sum_{i=2}^{397} \beta_i D_i \; + 0.00069 \; CR4 \qquad\qquad \textbf{(9.A7)}$$
$$(F = 2.9)^a \qquad (t = 1.4)^c$$
$$+ \; 0.00094 \; FED - 0.0000045 \; SALES$$
$$(t = 1.9)^b \qquad\qquad (t = -0.48)$$

F-value $= 3.1^a$, $R^2 = 0.47$, degrees of freedom $= 1,419$

$$SYST = \quad 0.27 \quad + 0.0023 \; CR4 + 0.0033 \; FED \qquad \textbf{(9.A8)}$$
$$(t = 12)^a \quad (t = 4.5)^a \qquad (t = 7.0)^a$$
$$- \; 0.000013 \; SALES$$
$$(t = -1.4)^c$$

F-value $= 27^a$, $R^2 = 0.043$, degrees of freedom $= 1,815$

Finally, the small negative effect associated with firm size does appear to be independent of the size of the line of business. Controlling for the 237 industry effects, *DIVERS*, *FED* and *SALES*, the coefficient on *FIRMSIZE* is again negative with a *t*-ratio of -1.26. Against the null hypothesis of a zero coefficient, the probability of a *t*-ratio that low or lower is 0.10435.

Notes

1. See Scherer (1984), Glennan (1967), Mansfield (1968) and Mansfield *et al.* (1977a) for important descriptions of the trial and error process of development.
2. When $Z = C$, the distribution of the sample space over x is crudely bell-shaped with relatively high degrees of component gestalt (low x outcomes) less numerous than relatively low degrees (high x outcomes). The text discusses the case relevant for an industrial R&D problem, where Z is much larger than C and most of the sample space is concentrated at $x = C$.

3. See Scherer (1984, pp. 222–38), Scott (1984), Levin *et al.* (1985) and Cohen *et al.* (1987).

4. Although to estimate the firm effects in the following specifications the effects were absorbed to reduce the size of the $X'X$ matrix, the F-values for firm effects (as well as the industry effects) are for the effects when fitted last (see Scott, 1984, p. 244). Significance levels for a one-tailed test are noted as $a = 0.0001$ and a^* is recorded for the F-value which was calculated in Scott (1984, p. 244).

5. Note that although there are 238 industries, 234 rather than 237 industry dummies are controlled here with both firm and industry effects in the model. Consider the special cases where *in the 1,819 observation sample* an industry and a company dummy coincide because a single-LB company is the sole producer in its industry category. Then, the number of industry dummies is reduced until no linearly dependent columns are in the X matrix.

6. Scott and Pascoe (1986) provide a more formal statement of the fact that the effects of firm (industry) level variables are nested within the effects of the firm (industry) dummies.

7. However, once seller concentration is controlled, but without control for other industry variables and without control for firm effects, systems developments decline very slightly with LB sales. Further, the small negative effect associated with firm size does appear to be independent of the size of the line of business.

8. We have in effect controlled for all relevant firm-level variables, any relevant powers of those variables, and their interactions, as well as all relevant industry-level variables, their powers and their interactions, and all relevant group-level variables, their powers and their interactions, assuming only that the true model is linear in the parameters, although not linear in the variables. The linear algebra and the Gauss–Markov theorem imply that the expected value of the intercept is the weighted sum of the entire set of coefficients for the true variables characterizing firms and industries, with the weight of each coefficient being the value its associated variable takes for the base firm or industry. The expected value of the coefficient of the dummy variable for each remaining industry (or firm) is the difference from the intercept of the weighted sum of the entire set of coefficients for the true variables characterizing industries (or firms), with the weight of each coefficient being the value its associated variable takes for that industry (or firm). Not all LBs are in a group, but for those that are, the coefficient of the dummy variable of any relevant group provides an additional effect that is the weighted sum of the entire set of coefficients for the true variables characterizing groups, with the weight of each coefficient being the value its associated variable takes for that group.

9. See the references in Note 3.

10. Mansfield *et al.* (1977b, pp. 236–8) find in a sample of industrial innovations that the gap between social and private rates of return is typically positive and increases, *ceteris paribus*, with the significance of the innovation. They measure an innovation's significance by its annual net social benefits. Society might prefer having a choice among innovations meeting a particular need, yet the monopolist might not appropriate the value such choice provides. If consequently the monopolist under-invests in diversity, competitors' extra incentive to pursue diverse strategies could be in the social interest.

150 *Scott*

References

Baldwin, W. L. and J. T. Scott, *Market Structure and Technological Change*, (New York: Harwood Academic Publishers, 1987).
Barzel, Y., "Optimal timing of innovations", *Review of Economics and Statistics*, vol. 50 (1968), pp. 348–55.
Chamberlin, E. H., "Duopoly: Value where sellers are few", *Quarterly Journal of Economics*, vol. 43 (November 1929), pp. 63–100.
Chamberlin, E. H., *The Theory of Monopolistic Competition* (Cambridge, Mass.: Harvard University Press, 1933).
Cohen, W. M., R. C. Levin and D. C. Mowery, "Firm size and R&D intensity: A re-examination", *Journal of Industrial Economics*, vol. 35 (1987), pp. 543–65.
Evenson, R. E. and Y. Kislev, "A stochastic model of applied research", *Journal of Political Economy*, vol. 84 (1976), pp. 265–81.
Glennan, T. K., Jr, "Issues in the choice of development policies" in *Strategy for R&D: Studies in the Microeconomics of Development*, edited by T. Marschak, T. K. Glennan, Jr and R. Summers (New York: Springer-Verlag for the Rand Corporation, 1967).
Levin, R. C., W. M. Cohen and D. C. Mowery, "R&D appropriability, opportunity and market structure: New evidence on some Schumpeterian hypotheses", *American Economic Review*, vol. 75 (1985), pp. 20–4.
Mansfield, E., *Industrial Research and Technological Innovation* (New York: W. W. Norton, 1968).
Mansfield, E., J. Rapoport, A. Romeo, E. Villani, S. Wagner and F. Husic, *The Production and Application of New Industrial Technologies* (New York: W. W. Norton, 1977a).
Mansfield, E., J. Rapoport, A. Romeo, S. Wagner and G. Beardsley, "Social and private rates of return from industrial innovations", *Quarterly Journal of Economics*, vol. 91 (1977b), pp. 221–40.
Nelson, R. R., "Uncertainty, learning and the economics of parallel research and development efforts", *Review of Economics and Statistics*, vol. 43 (1961), pp. 351–64.
Nelson, R. R., "The role of knowledge in R&D efficiency", *Quarterly Journal of Economics*, vol. 97 (1982), pp. 453–70.
Osborne, D. K., "Cartel problems", *American Economic Review*, vol. 66 (1976), pp. 835–44.
Scherer, F. M., *Industrial Market Structure and Economic Performance*, 2nd edition (Chicago: Rand McNally, 1980).
Scherer, F. M., "The propensity to patent", *International Journal of Industrial Organization*, vol. 1 (1983), pp. 107–28.
Scherer, F. M., *Innovation and Growth: Schumpeterian Perspectives* (Cambridge: MIT Press, 1984).
Schumpeter, J. A., *Capitalism, Socialism and Democracy* (New York: Harper, 1942).
Scott, J. T., "Firm versus industry variability in R&D intensity", in *R&D, Patents and Productivity*, edited by Zvi Griliches (Chicago: University of Chicago Press for the National Bureau of Economic Research, 1984), Chapter 10, pp. 233–45.
Scott, J. T., "Diversification versus cooperation in R&D investment", *Managerial and Decision Economics*, vol. 9 (1988), pp. 173–86.

Scott, J. T. and G. Pascoe, "Beyond firm and industry effects on profitability in imperfect markets", *Review of Economics and Statistics*, vol. 68 (1986), pp. 284–92.

Scott, J. T. and G. Pascoe, "Purposive diversification of R&D in manufacturing", *The Journal of Industrial Economics*, vol. 36 (1987), pp. 193–205.

10

Firm size, growth and innovation: Some evidence from West Germany

Felix R. FitzRoy and Kornelius Kraft

10.1 Introduction

This chapter provides some new evidence on two related areas of industrial organization, where traditional views have been changing rapidly in response to recent research findings.

The traditional wisdom on the relationship between firm size and growth was summarized by Gibrat's law of constant proportional growth, independent of size. This "law" seems, however, to be inconsistent with a large body of recent research in the United States, summarized by Acs and Audretsch (1990a). Small firms, particularly in manufacturing, have been growing faster and creating more jobs than large firms in recent years. Evans (1987a, b) found that firm growth decreases both with size and age of the firm, and related findings for Italy have been obtained by several authors.

Entry and exit of firms complicate the problems of sample selection. A stylized fact which appears to hold across countries is that most new start-ups fail fairly soon, but a small proportion either survive (and remain small), while a still smaller remnant grow rapidly (Brock and Evans, 1989). Large and small firms may grow by merger or acquisition, or themselves be acquired and sometimes dismembered (Storey and Johnson, 1987).

West Germany (FRG) represents a striking contrast to experience in the United States and Italy. As Acs and Audretsch (1990a, p. 17) summarize, "while there has been a dramatic drift in the share of manufacturing sales and employment accounted by small firms in the US, no such trend is apparent in the FRG". This conclusion contradicts several micro-economic studies of firm size and employment change in Germany discussed by Brunowsky (1988). According to these studies, small firms grew much faster than large ones in the last decade, which seems to contradict the lack of change in FRG small firm shares mentioned above. Dunne and Hughes (1989) find a small increase in FRG small firm employment. This work is descriptive rather than based on rigorous econometric analysis, and the sample selection bias that has plagued many studies in this area may help to explain the contradiction, which in any case poses an interesting puzzle for future research.

Our contribution here is to provide an econometric study of the size–growth relation in a small panel of firm-level data in the FRG metalworking industry. The sample is not representative, but is constant over time (without entry or exit) and contains information on innovation, incentives and human capital which is not usually available or related to growth performance.

In the Schumpeterian tradition, most economic studies of innovation have been concerned with market structure and firm size, and used cross-sectional data (Cohen and Levin, 1989). Only recently has it been shown that large firms are actually *less* innovative than small ones according to various measures in some industries (Acs and Audretsch, 1990a, b). Although innovation is widely believed to be an important reason for the faster growth of successful firms, there seems to be very little statistical evidence (as distinct from case studies) on this point.[1] The main innovation in our present study is thus to combine a measure of product innovation with size, age, human capital and other variables to explain growth performance.

In the next section we provide a detailed description of the data, and in Section 10.3 the key hypotheses and the econometric results for various alternative specifications of our basic model are presented. The relevance of these and related results for industrial policy is discussed in the conclusion.

10.2 Description of the data

Our unpublished data for fifty-one firms in the metalworking sector for 1977 and 1979 were collected by us in extended interviews and subsequent questionnaire responses. There were nine firms with more than 1,000 employees, six with 500–1,000 and thirty-six with less than 500 employees. All variables are defined in Table 10.1, together with mean values and variances. Before discussing some of the variables in more detail, we note that the sample was selected to study the effects of profit-sharing, practised in about 50 per cent of the sample. Firms were not "randomly" chosen, nor can we claim that the sample is in any way

Table 10.1 Description of variables.

Definition	Mean value	Standard deviation
UNSKILL: Ratio of unskilled to skilled blue collar workers	2.17	3.51
EXP: Proportion of sales exported	31.2	24.2
CTOP: Share of capital held by top management	0.69	0.47
HERF: Concentration index = inverse of the number of main competitors	0.14	0.12
YOUNG: Dummy = 1 if firm was founded after 1972	0.04	0.20
NEM: Number of employees in 1977	625.8	954.1
HIGH: Proportion of workforce with higher education	0.05	0.03
INNO: Proportion of 1979 sales consisting of products introduced since 1974	44.65	26.57
PROSH: Profit share per employee (1000 DM)	0.41	0.79
G: Growth rate of sales from 1977–79	0.19	0.35

"representative". However, it is fairly homogeneous, and includes firm specific variables not usually available. There is no obvious relationship between key growth and innovation variables, and our selection procedure is based on the matching of profit-sharing and non-profit-sharing firms, and their willingness to supply unpublished data.

Most measures of innovative output (such as counting innovations) suffer to some extent from an inability to evaluate the economic importance of an "innovation". Our measure (*INNO*) is based on responses to the following question: "What proportion of 1979 sales consists of products which were substantially further developed or newly developed internally within the last 5 years?" While this question was designed to exclude trivial modifications, it obviously does not completely solve the problem.[2] Finally, it is worth emphasizing that input measures such as R&D expenditures are likely to be very unreliable measures of innovation for small firms where much innovative activity is not covered by formal R&D budgets (Kleinknecht, 1989).

Turning to our other variables we use two proxies for incentives, the first of which is *CTOP*, the capital ownership share held by top management. Profit-sharing pay for other employees (though generally only a small percentage of wage payments turned out to be strongly related to productivity and profitability in earlier work)[3] where a theory of group incentives affecting cooperation in the workplace was also developed. According to managerial and agency theories of the firm, ownership by top management should reduce managerial shirking and improve firm performance. On the other hand, based on our theories and evidence, profit shares for all employees should encourage cooperation and mutual monitoring to increase productivity. Recent discussions of efficiency wages and rent sharing also suggest that equity motives or "gift exchange" (Akerlof, 1982) may induce profit-sharing or premium wages in more successful firms.[4]

Turning to human capital variables, *HIGH* is a proxy for educational capital, defined as the proportion of the workforce with higher education (university degree or equivalent). *UNSKILL* is the ratio of unskilled to skilled production workers according to the official classification. *EXP* is the share of sales exported, often considered to be an indicator of product quality and thus presumably related to managerial or production skills.

Size of the firm is measured by sales or the total number employed, *NEM*. To control for the age of the firm, which is an important variable in all life cycle theories, we use a dummy variable, *YOUNG*, which takes the value 1 if the firm was founded after 1969. Product market competition is described by *HERF*, which is the inverse of the number of main competitors in all relevant markets given by respondent firms.

10.3   Hypotheses and results

Three main hypotheses can be tested with our data as described above. First, following Evans (1987a, b) we expect smaller and younger firms to be growing faster. Second, following a widespread belief that has been subject to surprisingly little formal testing, we expect more innovative firms to grow faster. Third, as Finseth (1988) and FitzRoy and Vaughan-Whitehead (1989) have shown, profit-sharing may be associated with more stable or rapid growth, though causality is not easy to determine.

At this point it should perhaps be emphasized that our methodology is close to that recommended by Schmalensee (1989). Since all our variables are endogenous, as is so often the case in industrial organization (particularly cross-sectional) studies, we cannot claim to identify a precisely specified econometric model based on conventional theoretical notions. Our more "modest" approach is rather to provide a relatively precise *description* of the salient regularities and correlations in the data.

Growth of sales seems to be a natural choice for the independent variable in the context of the literature referred to above, so we define

$$G_i = (SALES_{i79} - SALES_{i77})/SALES_{i77} \qquad (10.1)$$

where $SALES_{it}$ refers to sales of firm i in year t. The more usual logarithmic specification yielded qualitatively similar but statistically somewhat less satisfactory results, so we do not report the logarithmic version here.[5] The linear specification is just

$$G_i = \Sigma a_k X_{ki} + e_i \qquad (10.2)$$

where the X_{ki} are characteristics of firm i as described above and e_i is the error term.

In view of heteroskedasticity, estimation was by weighted least squares (WLS). Using total employment as the measure of firm size, estimates of Equation 10.2 are reported in Table 10.2. The innovation variable *INNO* was treated as an instrumental variable in the second (*IV*) column, but since we do not have good instruments for this essentially pre-determined variable, it loses significance. However, the coefficient does not decline, indicating an absence of simultaneity problems that is also supported by the time lag of five years involved in the definition of the *INNO* variable. We thus focus attention on the results in the first column.

The three main hypotheses are all confirmed in our sample. Larger firms display significantly slower growth (negative significant coefficient for *TEM*), while the age dummy *YOUNG* is positive significant, so younger firms do grow faster, controlling for employment. The innovation measure is also positive and (just) significant. The human capital and incentive results were less uniformly in accordance with

Table 10.2 Regressions of sales growth: coefficients (and *t*-statistics).

	WLS	*IV*-WLS
CONSTANT	−0.32	−0.57
	(−1.67)	(1.80)
UNSKILL	0.02	0.02
	(1.32)	(1.39)
EXP	0.001	0.0004
	(0.44)	(0.14)
CTOP	0.07	0.12
	(0.58)	(0.98)
HERF	−0.46	−0.90
	(−0.46)	(−0.90)
YOUNG	0.82	0.65
	(3.88)	(1.86)
NEM	−0.0002	−0.0001
	(−3.13)	(−2.73)
HIGH	5.46	6.14
	(2.64)	(2.76)
INNO	0.004	0.009
	(1.89)	(1.23)
PROSH	0.21	0.20
	(3.59)	(2.97)
R^2	0.79	

expectations. Both *UNSKILL* and *CTOP* were insignificant. However, higher education and profit-sharing were both strongly positively related to growth as theory suggests. Finally, the market competition (*HERF*) and export share variables were quite insignificant. In unreported regressions we also used *SALES* as an independent variable to measure firm size, and obtained practically identical results.

10.4 Policy conclusions

Our results from a sample of German metalworking firms confirm recent findings with US data, that smaller and younger firms tend to grow faster. Our product innovation measure was also positively related to growth, a frequently assumed but seldom rigorously tested relationship. Higher education and profit-sharing were also associated with more rapid sales growth in our sample.

These results, while not of course in any way representative for the FRG, are of some interest inasmuch as they do lend support to the growing literature which is critical of official European policy towards large firms. As authors such as Acs and Audretsch (1990a), Adams and Brock (1988) and FitzRoy (1989, 1990) have emphasized from varying perspectives, an increasing number of research findings cast strong doubt on traditional assumptions of generally increasing returns (to scale and to R&D) and benefits from conglomerate diversification and merger.

On the other hand, European policy-makers and even many economists still extol the benefits – and indeed the necessity – of large organizations in order to compete effectively in international markets. These beliefs are matched by massive direct and indirect subsidies for large firms, and many legal and institutional obstacles for small firms attempting to obtain grants or break into quasi-monopolistic markets for government contracts. As 1992 approaches, merger and acquisition activity is increasing rapidly in Europe, usually with official support even in the absence of obvious synergy prospects such as in the case of Daimler–Benz and Messerschmitt-Bölkow-Blohm (MBB). The strong relationship between firm size and top managerial compensation found in German corporations by FitzRoy and Schwalbach (1989) matches US and UK findings, together with weak pay–performance relationships, and suggests powerful managerial incentives for wasteful growth and diversification (Jensen, 1989).

More research on the causes and effects of individual firm growth is clearly needed before current policies and official faith in scale economies are likely to be revised.

Notes

1. Mansfield (1962) and Gort and Klepper (1982) represent important, pioneering exceptions which illustrate the data problems that face researchers in this area. In their excellent survey, Cohen and Levin (1989) note "the profession's primitive understanding of the determination of the size and growth of firms" (p. 1070).
2. Patent counts are still widely used as measure of innovation, but were shown to be unsatisfactory by Gort and Klepper (1982); see also Cohen and Levin (1989).
3. See FitzRoy and Kraft (1986, 1987).
4. Wages showed little variation in our sample. Profit shares were also uncorrelated with wages.
5. Employment growth yielded the same sign pattern, but the fit was worse, perhaps due to neglect of capital–labor substitution possibilities.

References

Acs, Z. J. and D. B. Audretsch, "Small firms in the 1990s" in *The Economics of Small Firms: A European Challange* Boston: Kluwer Academic Publishers, 1990a.

Acs, Z. J. and D. B. Audretsch, *Innovation and Small Firms*, Cambridge: MIT Press, 1990b.

Adams, W. and J. W. Brock, "The bigness mystique and the merger policy debate: An international perspective", *Northwestern Journal of International Law and Business* vol. 9 (1988), pp. 1–45.

Akerlof, G., "Labor contracts as partial gift exchange", *Quarterly Journal of Economics* vol. 97 (1982), pp. 573–69.

Brock, W. A. and D. S. Evans, "Small business economics", *Small Business Economics* vol. 1 (1989), pp. 7–20.

Brunowsky, R., *Das Ende der Arbeitslosigkeit*, München Piper Verlag: 1988.

Cohen, W. M. and R. C. Levin, "Empirical studies of innovation and market structure" in R. Schmalensee and R. D. Willig (eds), *Handbook of Industrial Organization* vol. 2, Amsterdam: North Holland, 1989.

Dunne, P. and A. Hughes, *Small Businesses: An Analysis of Recent Trends in Their Relative Importance and Growth Performance in the UK with some European Comparisons* Cambridge: Small Business Research Centre, Dept of Applied Economics, Cambridge University, 1989.

Evans, D. S., "The relationship between firm growth, size and age: Estimates for 100 manufacturing industries", *Journal of Industrial Economics* vol. 35 (June 1987a), pp. 56–81.

Evans, D. S., "Tests of alternative theories of firm growth", *Journal of Political Economy* vol. 95 (August 1987b), pp. 657–74.

Finseth, E., "The employment behavior of profit-sharing firms", Honors Thesis, Dept of Economics, Harvard University, 1988.

FitzRoy, F. R., "Wage structures, employment problems and economic policy", Brussels: Commission of the European Communities, 1989.

FitzRoy, F. R., "Employment, entrepreneurship and 1992", *Small Business Economics* vol. 2 (January 1990), pp. 11–24.

FitzRoy, F. R. and K. Kraft, "Profitability and profit-sharing", *Journal of Industrial Economics* vol. 34 (December 1986), pp. 113–30.

FitzRoy, F. R. and K. Kraft, "Cooperation, productivity and profit sharing", *Quarterly Journal of Economics* vol. 102 (February 1987), pp. 23–36.

FitzRoy, F. R. and J. Schwalbach, "Managerial compensation and firm performance: Evidence from West Germany", mimeo, Berlin: Wissenschaftszentrum Berlin für Sozialforschung.

FitzRoy, F. R. and D. Vaughan-Whitehead, "Efficiency wages, employment and profit sharing in French firms", mimeo, Berlin: Wissenschaftszentrum Berlin für Sozialforschung, 1989.

Gort, M. and S. Klepper, "Time paths in the diffusion of product innovations", *Economic Journal* vol. 92 (1982) pp. 630–53.

Jensen, M., "The eclipse of the public corporation", *Harvard Business Review* vol. 92 (September–October 1989), pp. 61–75.

Kleinknecht, A., "Firm size and innovation: Observations in Dutch manufacturing industry", *Small Business Economics* vol. 1 (1989), pp. 215–22.

Mansfield, E., "Entry, Gibrat's law, innovation and the growth of the firms", *American Economic Review* vol. 52 (1962) pp. 1023–51.

Schmalensee, R., "Interindustry studies of structure and performance" in R. Schmalensee and R. D. Willig (eds), *Handbook of Industrial Organization*, vol. 2, Amsterdam: North Holland, 1989.

Storey, D. J. and S. Johnson, *Job Generation and Labour Market Change*, London: Macmillan, 1987.

11

Advertising, innovation and market structure: A comparison of the United States of America and the Federal Republic of Germany*

J.-Matthias Graf von der Schulenburg and
Joachim Wagner

11.1 Introduction

To obtain a better insight on the inter-relationship of advertising, innovation and concentration we will compare empirical results derived from data of US industries and West German industries. To enable an international comparison, we have tried to employ comparable variables of both countries and to estimate similar specified models for both countries. Although some of the variables can only be interpreted as proxies, the comparison of the results for both countries may lead to a better understanding of the interdependence between market structure, advertising intensity and innovation. If our model leads to similar results for both countries it shows the robustness of the model employed and supports the specification of the model. If, however, the results differ for both countries this indicates that either the model is misspecified or significant differences exist between both countries which have to be explained, e.g. differences in industrial policy, market characteristics, corporate cultures and institutional regulations. The comparative study will also reflect that the US market is much larger than the German one, less export oriented and less import dependent. In any case, an international comparison is always helpful for testing a model and receiving better insights in the inter-relationship of economic variables.

* An earlier version of this chapter was presented at the Conference on Innovation and Technological Change: An International Comparison at Science Centre Berlin, August 10–11, 1989. We thank Paul Geroski and Frederic M. Scherer for helpful comments. The usual disclaimer applies.

Why are we interested in the inter-relationship of advertising and innovation with respect to market structure? In a world of perfect information there would be no advertising and no need to invest in innovative activites. Both phenomena are results of imperfect information. Consumer ignorance gives an incentive to suppliers to spend advertising expenditures. The lack of information about the commodity space and production possibilities is the reason that product and process innovations are possible. If however, information about inventions is available to everyone at zero cost, no one has an incentive to spend resources for inventions and innovations. If information is not available on zero cost or may not be employed by others due to patent or copyright laws, innovations lead to some monopoly power and extra profit. Thus, whether information produced through innovation is private or public should have an effect on the propensity for innovation.

Obviously advertising and innovation have some characteristics in common. They are the results of imperfect information. They both produce information which should be apparent for inventions and innovations, and which was pointed out by Telser (1964) for advertising expenditure. Innovation as well as advertising are frequent phenomena in concentrated markets and work as barriers to entry for new market entrants.[1] Both increase the monopoly power of a firm, and both can be treated as special forms of investments because the benefits are in future periods and uncertain.

Despite those common characteristics, studies on the inter-relationship of advertising and innovation are rare.[2] This has a number of reasons. Traditionally advertising was described by economists as a wasteful activity which weakens the efficiency of price competition.[3] Only in newer economic studies analyzing markets with imperfect competition and information was advertising treated as a form of non-price competition and as an entry barrier determining the market structure.[4] Orr (1974), Khemani and Shapiro (1986) and Audretsch and Schulenburg (1990) found advertising to be a significant barrier to entry, while Duetsch (1984) and Acs and Audretsch (1987) found no empirical evidence for the hypothesis that advertising intensity deters entry for small firms, and therefore, has no significant influence on the market structure.

Economic studies on the determinants of innovation have a long tradition since the classical publications by Arrow (1962) and Schumpeter (1950). On the basis of Schumpeter's hypothesis that market concentration has a significant effect on innovation, a substantial literature relating market structure to technical change has evolved.[5]

In most economic studies, advertising intensity is treated as independent of market structure while innovation is treated as being dependent. In addition, economists have paid very little attention to the inter-relationship between advertising and innovation. In this chapter, we study the relationship of two phenomena arising from imperfect information, namely advertising and innovation. Both economic activities seem to depend upon the market structure, particularly upon the concentration ratio. In contrast to many other studies we treat concentration and innovation in a simultaneous framework as well, because there are good theoretical arguments for the hypothesis developed in "the new industrial organization"[6] which considers the characteristics of technology, including the technology of invention and innovation to be key factors for explaining the structure of markets. Similar arguments call for an endogenous treatment of advertisment with respect to innovation. If advertising is done to provide information on new products, the advertising intensity should be determined by the innovative activities of an industry. On the other hand, a high level of advertising in an industry works as an entry barrier and, thus, has a modifying effect on the incentives to innovate. Therefore, advertising should also be treated as an endogenous variable to control for a simultaneity bias, causing misleading interpretation about causal relationships.

11.2   The interrelationship between advertising, innovation and market structure

Advertising is traditionally employed as an independent variable with respect to market structure. However, the advertising expenditures of companies most likely are not independent of the structure of the market, the features of the commodities sold and the behavior of competitors.[7] For a theoretical treatment of this problem, let us distinguish two types of advertising, namely innovation advertising, i.e. advertising providing information about a new product or a new technology, and product differentiation advertising for brand commodities which are close substitutes.

Innovation advertising is expected in innovative industries where consumers need information about new products. Product differentiation advertising is instead expected in oligopolistic markets of experience goods and search goods. In the case of innovation advertising we expect a positive influence of innovative activities on advertising intensity.

Product differentiation advertising is not uninformative. In Butters' (1977) model, for instance, advertising is done only by companies producing similar goods to provide price information. But even if advertising contains no product information and is purely persuasive, i.e. merely affects the consumers' preferences for certain brands as in the model by Dixit and Norman (1978), it may work as a signal for the product quality of an experience good.[8] Such advertising activities are typical in markets with monopolistic competition, where brands are known to the consumer and price competition and advertising are employed by companies at the same time. We expect, therefore, that market concentration has a significant positive influence on the advertising intensity of an industry.[9]

Advertising exists only in markets with imperfect information. As market studies show,[10] private consumers are relatively poorly informed about goods and quality levels available in a market and about prices. Corporate and professional customers are much better informed and spent more in gathering information and comparing offers by different suppliers. We would therefore expect that the advertising expenditures are higher in private consumption-oriented industries. The beverage, liquor, toiletry and detergent industries are especially known for their costly and massive advertising campaigns to attract the attention of the ordinary consumer.

Concerning the variables influencing the innovation intensity, the economic literature suggests that concentration has a positive impact on innovation. As Schumpeter (1950) has argued, markets with imperfect competition are most conducive to innovation, because in imperfect markets firms are able to capture the economic rents accruing from innovations. However, empirical support for Schumpeter's hypothesis remains weak,[11] and some empirical studies have even found a negative relationship between research and development expenditures and concentration.[12]

If the incentives to spend resources on innovative activities are higher in markets characterized by imperfect competition, as claimed by Schumpeter (1950), then the advertising intensity should also have a positive effect on innovation.[13] Advertising can make the demand for a firm's brand more price inelastic and also decrease the cross-price elasticities of demand with respect to rival brands.[14] Since advertising is also done to make the buyer aware of a high quality brand, there is an incentive to invest constantly in innovations; otherwise, advertising expenditure will have a smaller return. Consumers will quickly learn that brands with high advertising budgets do not necessarily provide exceptional quality.

Recently, several models have been developed in literature arguing that unions have an influence on the innovation intensity of an industry. Connolly *et al.* (1986) developed a model where unions act as a distortionary tax on the returns from investment in intangible capital.[15] If those investments such as innovations lead to higher rents, these rents will be partly absorbed by higher wages and fringe benefits, lower working time, etc. The hypothesis that unions are capturing rents on returns from intangible capital is supported by the findings of lower stock market values on R&D investments by unionized firms as compared to non-unionized firms.

However, in older German economic discussions one can find a different argument. Unions fight for higher real wages, which forces firms to invest in process innovation in order to increase the substitution of capital for labor, and to invest in product innovation to increase the profitability of labor. The wage push hypothesis implies a positive influence of union power on innovation. Which of these contradicting hypotheses holds depends also on the way in which contracts between unions and employers are written. If contracts allow a quick adjustment of wages and working conditions, then the rent seeking hypothesis is more plausible because employers can always expect that some share of their innovation rents will be absorbed by unions. So, empirical studies are needed to clarify the relationship between union participation and innovation. Thus far, only a handful of studies have examined this relationship empirically.[15] Audretsch and Schulenburg (1990) found a significant negative relationship for the United States. The significance of the estimated coefficient disappeared, however, when the relationship of union participation, innovation per sales and concentration were tested in a simultaneous equations model.

Three additional variables are included in the current model to explain the variation in innovation activity: human capital, firm size and research and development (R&D) expenditure. Human capital and R&D expenditure should have a positive impact on the innovation of a firm. However, one should note that in some industries a large proportion of R&D expenditure is spent for imitation rather than for innovation. Thus, the expenditure for "real" research and development may be less than the R&D budget. In the pharmaceutical industry, for instance, it is common to include expenditure for "scientific marketing" in the R&D budget and allowances to physicians. According to the Schumpeterian hypothesis, industries with mainly large firms should be more innovative than those with mainly small firms.[16] This positive relationship is also found by Acs and Audretsch (1988) employing US data. However, the study by Acs and Audretsch did not support the Schumpeterian hypothesis with respect to the relationship of concentration and innovation. They found that industry innovation tends to decrease as the level of concentration rises.

The third key variable in our analysis is the market structure as represented by the concentration ratio. Stiglitz (1986) has argued that non-convexity and irreversibilities, or sunk cost, are decisive in determining concentration. He observes that, while many different technologies fulfill one of these characteristics, innovation and product differentiation strategies typically match both non-convexity and irreversibilities.[17] Therefore, we expect that market concentration is positively influenced by innovation and advertising efforts. Scale economies, the wage level and the capital intensity of an industry influence the market structure because they impede the market entry of new firms.[18] Therefore, capital intensity and the wage level are also expected to be positively correlated with concentration. The same is expected for import tariff barriers. Tariffs protecting an industry normally decrease competition within that industry and lead to an increase of market concentration. This might, however, cause a counter reaction. Krugman argues, "protection by initially generating monopoly rents, generates excessive entry and thus leads to inefficiently small scale production. This proposition, originally proposed by Eastman and Stykolt, is backed by substantial evidence, and has been modelled by Dixit and Norman".[19] Therefore concentration might be less than expected with respect to entry barriers, and the same is true for profits.

Table 11.1 summarizes the various theoretical hypotheses of the determinants of advertising and innovation intensities and concentration, and the expected signs of the coefficients.[20]

11.3 Econometric study

The model outlined in Table 11.1 was estimated using data for twenty-nine industries (two-digit level SYPRO) in Germany and 247 industries (four-digit level of SIC) in the United States. Table 11.2 provides detailed information on the variables. Most of the variables are standard, but some of them deserve comment. First, the innovation measures both for Germany and for the United States are output

Table 11.1 General structure of the model.

Equation no.	Endogenous variable	Exogenous variables (theoretically expected sign given in brackets)		
1	Advertisting	Concentration (+)	Innovation (+)	Private consumption (+)
2	Innovation	Concentration (+)	Advertising (+)	Human capital (+)
		Union power (?)	Firm size (+)	R&D (+)
3	Concentration	Innovation (+)	Advertising (+)	Capital intensity (+)
		Wage level (+)	Tariffs (+)	

measures. The US innovation measure is taken from a database
released by the US Small Business Administration. It consists of over
8,000 innovations which were identified from more than 100 technology,
engineering and trade journals. The innovation measure in the German
data set is the percentage of shipments of those products which were
introduced recently into the market and are still in the entry phase.
Second, no advertising-to-sales ratio data are available for Germany,
and, therefore, we apply a proxy based on figures for employees in
marketing activities. Third, due to peculiarities of the German union
system, no data on union density at the industry level are published,
because most unions cover more than one industry and in most
industries employees can be a member of one of several unions. We,
therefore, are forced to use a proxy variable – the percentage of female
employees which are known to be much less unionized than men.
However, we admit that the estimated coefficients of these proxies
should be interpreted with caution. Furthermore, it should be obvious
from the definitions given in Table 11.2 that only one variable (i.e. the

Table 11.2 Empirical measures used for Germany and the United States.

Variable	FRG data	US data
Innovation	Percentage of shipments due to products recently introduced in the market and still in the market entry phase	Number of innovations per employee
Concentration	Herfindahl index	Four-firm concentration ratio
Advertising	Percentage of employees in marketing activities	Advertising to sales ratio
Human capital	Percentage of employees with a university degree	Percentage of professional and kindred workers
Union influence	Percentage of female employees (proxy; negatively correlated with unobserved union density)	Union density
Firm size	Number of employees per firm	Dummy for large firm industries
R&D	Percentage of revenues spent on R&D	Scientists in R&D as a percentage of total employment
Capital intensity	Value of capital stock per man hour	Capital stock per employee
Wage level	Wage per man hour	Wage per man hour
Tariffs	Total protection in Germany (effective rate of protection plus subsidies)	US simple average tariffs
Private consumption	Percentage of shipments delivered for private consumption	Dummy for consumption industries

Note: See also the Data appendix at the end of the chapter.

wage level) is defined exactly identical in the US and the FRG data sets. Although we tried to specify the variables as identical as possible, one should, therefore, keep these differences as regards the definitions in mind when comparing the results internationally.

As a first step, the three equations of the model were estimated separately using ordinary least squares (OLS). It is well known that the results of OLS can be heavily influenced by a small number of unusual observations ("outliers").[21] We, therefore, have applied a mean shift outlier test to detect such industries by successively adding dummy variables for each industry (one at a time) to the equations and testing the maximum t-values of the coefficients.[22] In a second step, we augmented all equations by adding dummies for the outliers detected. On the one hand, this can be viewed as one way to specify a "robust"

Table 11.3 Innovation – equation for Germany.

Exogenous variable	Method	OLS	OLS	2SLS	3SLS
Constant	β	8.3857	3.946	3.397	4.225
	$\|t\|$	2.50*	1.80	1.70	2.14*
	ATV	1.97	1.95		
Concentration	β	0.0319	−0.005	−0.022	−0.018
	$\|t\|$	0.87	0.23	0.79	0.65
	ATV	0.88	0.32		
Advertising	β	−0.8054	−0.721	−1.007	−0.992
	$\|t\|$	0.39	0.57	0.91	0.90
	ATV	1.79	2.48*		
Human capital	β	−0.4798	0.009	0.241	0.146
	$\|t\|$	0.72	0.02	0.54	0.33
	ATV	0.85	0.03		
Union influence[a]	β	−0.0273	−0.102	−0.108	−0.086
	$\|t\|$	0.34	2.01	2.45*	1.98
	ATV	0.36	2.15*		
Firm size	β	0.00001	0.001	0.002	0.002
	$\|t\|$	0.00	0.69	1.06	1.13
	ATV	0.01	1.38		
R&D	β	0.1224	0.443	0.530	0.498
	$\|t\|$	0.24	1.37	1.83	1.73
	ATV	0.30	1.47		
D10 (Shipbuilding)	β		28.105	28.983	28.666
	$\|t\|$		6.06**	7.04**	6.99**
	ATV		17.35**		
R^2		0.059	0.657		
$\bar{R}^2$		−0.198	0.543		
N of cases		29	29	29	29

Notes: β = estimated regression coefficient
$\|t\|$ = absolute t-value
ATV = heteroskedasticity-consistent t-value
*(**) = significant at a 5 per cent (1 per cent) level (two-tailed test)
For a definition of variables see text.
[a]The union proxy is *negatively* correlated with the union density; the sign of the estimated coefficient, therefore, is reversed to allow an easy comparison with the results from the US equation.

model which fits the bulk of the data well without giving too much weight to extreme cases. On the other hand, adding dummies for outliers is one rather crude way to deal with inter-industry variations in, e.g. opportunities to innovate due to factors not explicitly modeled in our equations; we will return to this point in our conclusions. It should be added that we do not intend to interpret the coefficients of the outliers, although rather plausible stories could be told at least for some of them in the US equation. Furthermore, since heteroskedasticity is often present in cross-sectional analysis, we have computed heteroskedasticity-consistent t-values ($ATVs$) as well.[23]

The results are reported in Columns 1 and 2 of Tables 11.3–11.5 for Germany, and Tables 11.6–11.8 for the United States. In all but one equation, outliers were present, and adding the outlier dummies often changed the results drastically.

Focusing on the results of the OLS estimations for the models with the outlier dummies and the case of Germany it can be seen from Column 2 of Table 11.3 that there is no statistically significant influence

Table 11.4 Concentration – equation for Germany.

Exogenous variable	Method	OLS	OLS	2SLS	3SLS
Constant	β	−390.997	−266.956	−253.92	−237.396
	|t|	5.79**	4.48**	4.73**	4.65**
	ATV	4.15**	3.89**		
Innovation	β	0.656	0.374	0.967	0.474
	|t|	0.35	0.26	0.69	0.34
	ATV	0.68	0.47		
Advertising	β	0.331	139.447	157.154	165.689
	|t|	0.02	3.89**	4.49*	4.96**
	ATV	0.02	6.21**		
Capital intensity	β	−71.835	−14.776	−4.064	−15.477
	|t|	2.65*	0.60	0.17	0.71
	ATV	2.51*	0.75		
Wage level	β	21.176	12.583	11.221	10.874
	|t|	6.00**	3.71**	3.52**	3.62**
	ATV	4.11**	3.32**		
Tariffs	β	1.044	0.115	−0.008	0.095
	|t|	2.33*	0.28	0.02	0.28
	ATV	2.55*	0.52		
D24 (Printing)	β		−517.27	−571.113	−600.769
	|t|		4.18**	4.82**	5.28**
	ATV		7.34**		
R^2		0.675	0.819		
$\bar{R}^2$		0.604	0.769		
N of cases		29	29	29	29

Notes: β = estimated regression coefficient
\|t\| = absolute t-value
ATV = heteroskedasticity-consistent t-value
*(**) = significant at a 5 per cent (1 per cent) level (two-tailed test)
For a definition of variables see text.

Table 11.5 Advertising – equation for Germany.

Exogenous variable	Method	OLS	OLS	2SLS	3SLS
Constant	β	0.393	0.105	0.125	0.135
	$\mid t \mid$	1.54	1.78	2.27**	2.46*
	ATV	1.25	2.18*		
Concentration	β	0.002	0.001	0.001	0.002
	$\mid t \mid$	1.37	2.53**	2.85**	3.91**
	ATV	1.49	2.13*		
Innovation	β	−0.005	0.007	0.004	0.002
	$\mid t \mid$	0.22	1.37	0.77	0.32
	ATV	0.45	1.15		
Private	β	−0.001	0.004	0.004	0.003
consumption	$\mid t \mid$	0.11	2.48*	2.90**	2.72*
	ATV	0.15	3.42**		
D24 (Printing)	β		3.442	3.436	3.432
	$\mid t \mid$		21.02**	23.33**	23.36**
	ATV		99.55**		
D11 (Aircraft)	β		1.227	1.201	1.046
	$\mid t \mid$		5.43**	5.70**	5.27**
	ATV		6.27**		
R^2		0.078	0.958		
$\bar{R}^2$		−0.033	0.949		
N of cases		29	29	29	29

Notes: β = estimated regression coefficient
$\mid t \mid$ = absolute t-value
ATV = heteroskedasticity-consistent t-value
*(**) = significant at a 5 per cent (1 per cent) level (two-tailed test)
For a definition of variables see text.

of concentration, human capital, firm size and R&D on innovation. Advertising has a negative influence (which is significant if a test is based on the ATV), and the union influence is negative, too. Recall, however, that we were forced to apply a proxy for union participation of doubtful quality. In the concentration equation, all but the capital intensity variables have the theoretically expected positive signs, though the coefficients of innovation and tariffs are not significant at a 1 per cent level. In the advertising equation, concentration and private consumption have statistically significant positive coefficients as theoretically expected, while the influence of innovations turns out to be positive but insignificant. The results for Germany do not support the Schumpeterian hypothesis that the market structure has a significant impact on the innovation intensity. They also do not back the belief that large firms are more innovative than small ones. Altogether, innovation and concentration are less correlated in Germany than might be expected from the industrial organization literature. If we control for outliers, i.e. employ a dummy for the shipbuilding industry in the German innovation equation, union influence turns out to have a highly significant negative influence on innovations.

Table 11.6 Innovation – equation for the United States.

Exogenous variable	Method	OLS	OLS	2SLS	3SLS
Constant	β	0.1897	0.1705	0.182	0.190
	| t |	1.81	2.20*	2.20*	2.31*
	ATV	2.23*	2.53*		
Concentration	β	−0.003 28	−0.004 08	−0.005 7	−0.005 8
	| t |	2.28*	3.83**	2.33*	−2.40*
	ATV	2.28*	4.39**		
Advertising	β	29.726	40.186	50.06	50.36
	| t |	1.56	2.86**	2.85**	2.89**
	ATV	1.80	2.69**		
Human capital	β	6.312	3.417	3.595	3.672
	| t |	4.59**	3.17**	3.25**	3.35**
	ATV	2.89**	2.29*		
Union influence	β	−0.0045	−0.001 25	−0.000 57	−0.000 69
	| t |	2.67**	0.99	0.37	0.46
	ATV	2.38*	1.04		
Firm size	β	0.0155	−0.026 7	−0.035	−0.030
	| t |	0.18	0.41	0.54	0.47
	ATV	0.16	0.37		
R&D	β	−0.088 8	−0.001 2	−0.000 09	−0.004 4
	| t |	1.58	0.03	0.002	0.10
	ATV	1.07	0.02		
D3811	β		2.272	2.255	2.259
(Engineering and	| t |		6.87**	6.75**	6.81**
scientific	ATV		18.93**		
instruments)					
D3576 (Scales and	β		2.953	2.973	2.960
balances)	| t |		8.97**	8.96**	8.99**
	ATV		24.25**		
D2087 (Flavoring	β		2.605	2.663	2.637
extracts syrup)	| t |		8.07**	8.00**	7.99**
	ATV		66.12**		
D2843 (Surface	β		1.374	1.358	1.391
active agents)	| t |		4.28**	4.20**	4.34**
	ATV		32.14**		
D3573 (Computers)	β		1.187	1.182	1.173
	| t |		3.50**	3.46**	3.46**
	ATV		6.89**		
R^2		0.247	0.603		
$\bar{R}^2$		0.228	0.585		
N of cases		247	247	247	247

Notes: β = estimated regression coefficient
\| t \| = absolute t-value
ATV = heteroskedasticity-consistent t-value
*(**) = significant at a 5 per cent (1 per cent) level (two-tailed test)
For a definition of variables see text.

For the United States we found a positive (as expected theoretically) and statistically significant influence of advertising and human capital on innovation, while concentration has a significant negative coefficient. The estimated coefficients of union participation, firm size and R&D are

Table 11.7 Concentration – equation for the United States.

Exogenous variable	Method	OLS	2SLS	3SLS
Constant	β	−2.150	−1.832	−0.662
	|t|	0.26	0.22	0.08
	ATV	0.30		
Innovation	β	−3.575	1.717	1.639
	|t|	1.43	0.53	0.50
	ATV	1.37		
Advertising	β	2398.06	1979.9	2019.7
	|t|	2.94**	2.05*	2.08*
	ATV	3.46**		
Capital intensity	β	0.314	0.330	0.333
	|t|	4.28**	4.44**	4.48**
	ATV	3.98**		
Wage level	β	8.990	8.389	8.101
	|t|	4.01**	3.68**	3.55**
	ATV	3.96**		
Tariffs	β	0.781	0.803	0.767
	|t|	3.09**	3.14**	3.00**
	ATV	3.26**		
R^2		0.220		
$\bar{R}^2$		0.204		
N of cases		247	247	247

Notes: β = estimated regression coefficient
$|t|$ = absolute *t*-value
ATV = heteroskedasticity-consistent *t*-value
*(**) = significant at a 5 per cent (1 per cent) level (two-tailed test)
For a definition of variables see text.

negative and insignificant. In the concentration equation, all coefficients except the one for innovations have the theoretically expected (and statistically significant) positive signs. The same is true for the advertising equation, where, again, the coefficient for innovation is not significant at a conventional level. The negative influence of concentration on innovation is unexpected. This result, that more concentrated industries are less innovative than less concentrated ones, is in contradiction with the Schumpeterian hypothesis. It is, however, in line with the later view of Schumpeter and new theoretical findings.[24]

To take the interdependent relations between innovation, concentration and advertising into account, the three equations were estimated for both countries using two-stage least squares (2SLS) and three-stage least squares (3SLS). The results are reported in the last two columns of Tables 11.3–11.8. Before turning to the results of these simultaneous equation models, it should be noted that a Hausman (1978) test[25] rejected the null hypothesis that OLS is the correct method of estimation only for the advertising equation in the model for the United States (at a 5 per cent level). This might indicate that simultaneity does not play an important role in our model. On the other hand, according

Table 11.8   Advertising – equation for the United States.

Exogenous variable	Method	OLS	OLS	2SLS	3SLS
Constant	β	−0.00002	0.00017	0.00017	0.00009
	$\lvert t \rvert$	0.12	1.45	0.98	0.54
	ATV	0.15	1.44		
Concentration	β	0.00001	0.00001	0.000007	0.000009
	$\lvert t \rvert$	2.07*	2.74**	1.74	2.18*
	ATV	2.84**	2.44*		
Innovation	β	0.0004	0.00005	−0.00003	−0.00003
	$\lvert t \rvert$	2.28*	0.45	0.24	0.22
	ATV	1.55	0.55		
Private	β	0.0013	0.0008	0.0008	0.00079
consumption	$\lvert t \rvert$	7.01**	7.09**	6.69**	6.85**
	ATV	5.03**	6.08**		
D2085 (Liquor)	β		0.00797	0.00798	0.00788
	$\lvert t \rvert$		9.90**	9.88**	9.82**
	ATV		58.31**		
D2841 (Soaps,	β		0.00585	0.00585	0.00575
detergents)	$\lvert t \rvert$		7.25**	7.23**	7.16**
	ATV		39.56**		
D2842 (Polishes,	β		0.00554	0.00565	0.00570
sanitation)	$\lvert t \rvert$		6.78**	6.82**	6.93**
	ATV		31.89**		
D2844 (Toiletries)	β		0.01173	0.0118	0.0117
	$\lvert t \rvert$		14.45**	14.43**	14.41**
	ATV		76.38**		
D3421 (Cutlery,	β		0.00716	0.00716	0.00709
razor blades)	$\lvert t \rvert$		8.89**	8.87**	8.84**
	ATV		52.00**		
R^2		0.194	0.724		
$\bar{R}^2$		0.184	0.715		
N of cases		247	247	247	247

Notes: β = estimated regression coefficient
$\lvert t \rvert$ = absolute t-value
ATV = heteroskedasticity-consistent t-value
*(**) = significant at a 5 per cent (1 per cent) level (two-tailed test)
For a definition of variables see text.
[a]The union proxy is *negatively* correlated with the union density; the sign of the estimated coefficient, therefore, is reversed to allow an easy comparison with the results from the US equation.

to our tests, advertising seems to be endogenous in the concentration equation for Germany and the United States (and vice versa), and concentration seems to be endogenous in the innovation equation for the United States.

A look at the 2SLS and 3SLS results for Germany reveals no striking differences compared to the OLS results, and the same is true for the US model, although some coefficients differ in their order of magnitude and some insignificant ones even change their signs. To facilitate the international comparison[26] the signs of the results for the 3SLS estimations are reported again in Table 11.9 for Germany, and the

Table 11.9 A comparison of the model for Germany and the United States.

	Innovation		Concentration		Advertising	
Exogenous variable	FRG	US	FRG	US	FRG	US
Innovation			(+)	(+)	(+)	(−)
Concentration	(−)	−			+	+
Advertising	(−)	+	+	+		
Human capital	(+)	+				
Union influence	(−)	(−)				
Firm size	(+)	(−)				
R&D	(+)	(−)				
Capital intensity			(−)	+		
Wage level			+	+		
Tariffs			(+)	+		
Private consumption					+	+

Note: The comparison is based on the 3SLS results reported in Tables 11.3–11.8; a sign in brackets indicates a coefficient not significant at a 5 per cent level. The dummies are ignored for this comparison.

United States. The results for the advertising equations and the concentration equations turned out to be rather similar and in accordance with the theoretical expectations. Contrary to this, the results for innovations not only differ between the countries, but are different from the theoretically expected ones and often insignificant statistically, as well. One reason for these results might be the difference in the way innovations are measured for both countries. Furthermore, the rather "poor" results for Germany might indicate a wrong specification of the innovation equation.

11.4 Conclusion

What did we learn from our comparison of the United States and Germany as regards the relationships among advertising, innovation and concentration? In short, our results indicate that simultaneity matters, and that the determinants of concentration and advertising seem to be rather similar in both countries. Perhaps, this shows that our model is quite acceptable, because it "works" quite well when applied to different countries which are similar in some aspects (e.g. both are highly industrialized market economies) and different in others (e.g. Germany is much more integrated in the world markets than the United States). In both countries we found a significant positive relationship between advertising and concentration. Advertising works as an effective barrier to entry and is a phenomenon most characteristic for oligopolistic (concentrated) markets. Although we have employed a rather weak proxy for union influence in the German equations, both countries appear to be characterized by a negative influence of union power on innovations. However, this relationship is not significant in a statistical sense.

The innovation measures employed in our study are output-oriented and measure product innovation. Process and technology innovation, however, also plays a dominant role in modern industrial processes. It might be that our estimations would have produced stronger results if process innovations were included in our study. However, in many cases new products lead to changes of the production technology and vice versa, so that a high positive correlation between process and product innovation can be expected.

There remains much to be done in investigating the determinants of innovation in a framework similar to the one applied here. With regards to Germany, our list of caveats is long: we used data for a rather high level of aggregation and a proxy for the union influence which was less than perfect. Perhaps most important, aspects dealing with international economic linkages were nearly completely ignored. Taking account of the role of import pressure for innovations,[27] the role of innovations for international competitiveness[28] and the endogenous nature of protection[29] might help to improve the model using data at the industry level.

Furthermore, it has long been recognized that inter-industry differences in technological opportunity might play a key role in understanding the determinants of the variations in the degree of innovativeness between industries. Discussing inter-industry differences in patenting, holding firm sales constant, Scherer[30] argued twenty-five years ago that

(p)erhaps most important is a set of influences best described under the heading "technological opportunity". Technological opportunity in this context could relate partly to industry traditions or to demand conditions not manifested in mere sales volume, but it seems most likely to be associated with dynamic supply conditions dependent in turn upon the broad advance of scientific and technological knowledge.

One way to deal with these differences in technological opportunities consists in adding dummies for industries known (or supposed to be) traditional innovators (or non-innovators), and an example for this approach can be found in the contribution of Scherer.[31] It should be noted that our approach of adding dummy variables for outliers is in some way related to this technique, although we let a statistical procedure decide which industries are much different in the technological opportunities to innovate (or the "propensity to advertise", etc.) compared to the rest instead of referring to some kind of folklore, widely shared opinion, lasting experience or whatever.

A more satisfactory way to deal with unobserved inter-industry differences, which might not only be important in the explanation of innovations but may play a major role in determining advertising and concentration, too, consists in modeling these unobservables as fixed or random effects in a model applied to pooled cross-section time series data for industries.[32] Recently, Geroski[33] applied a fixed effects model to control for technological opportunity in a study of the links between market structure and innovation. However, we were unable to use panel data models here because we only have information on innovations in the United States for one year. Our future research for Germany will use pooled data for industries to control for unobservables like differences in technological opportunity.

A promising research perspective seems to be the potential use of data from a panel of firms which would allow for control of unobservable, firm-specific effects in a dynamic framework. We are currently planning to start such a study as part of an international research network. Hopefully, international comparisons using the database that will grow out of this project will be possible by the mid-1990s.

Data Appendix

The German data are for 1982/83 and were collected by J. Wagner. The US data were kindly supplied by Z. J. Acs and D. Audretsch (Wissenschaftszentrum für Sozialforschung Berlin). For a detailed description see Acs and Audretsch (1988) and Audretsch and Schulenburg (1990).

FRG data

All data are for the 29 two-digit SYPRO industries covering the manufacturing sector of Germany with the exception of foodprocessing and agricultural product-based industries.

1. Innovation. The data for the percentage of shipments due to

products recently introduced to the market are taken from unpublished data based on the innovation test conducted by the IFO Institute in Munich. The IFO Institute asked firms to specify what percentage of shipments are due to products in a certain stage of product life cycle, namely the entry stage, growth stage, stage where they have already reached the peak and shrinking stage. Innovation is measured as percentage of shipments of products in the market entry stage.

2. Concentration. The data for the Herfindahl index are taken from: Statistisches Bundesamt, Fachserie 4, Reihe 4.2.3, *Konzentrationsstatistische Daten für den Bergbau und das Verarbeitende Gewerbe sowie das Baugewerbe 1983 und 1984*, Wiesbaden: Kohlhammer 1986.

3. Advertising. The data for the percentage of employees in marketing activities are computed from unpublished data for the structure of employees supplied by Professor Franz-Josef Bade, University of Dortmund, which are based on the statistic of employees with social insurance (*Beschäftigtenstatistik*).

4. Human capital. The data for the percentage of employees with a university degree are computed from unpublished data supplied by Dr Ulrich Jung, Niedersächsisches Institut für Wirtschaftsforschung, which are based on the *Beschäftigtenstatistik*.

5. Union influence. The data for the percentage of female employees are computed from data supplied by Günther Schmid, Wissenschaftszentrum Berlin für Sozialforschung, taken from the "Regionaldatenbank Arbeitsmarkt" ("Regional data base for the labor market") and based on the *Beschäftigtenstatistik*.

6. Firm size. The number of employees per firm is computed from data for the number of employees taken from: Egon Baumgart, Rosemarie Mehl, Joachim Schintke, *Produktionsvolumen und -potential, Produktionsfaktoren des Bergbaus und des verarbeitenden Gewerbes in der Bundesrepublik Deutschland*, 26, Folge, Berlin: DIW 1984 (Table 18) and from data for the number of firms taken from: Statistisches Bundesamt, Fachserie 4, Reihe 4.2.3. (cf. above definition 2. Concentration).

7. R&D. The data for the percentage of shipments spent on R&D are taken from unpublished material of the "Stifterverband für die deutsche Wissenschaft" prepared for the Institute for Quantitative Economic Research, University of Hannover.

8. Capital intensity. The data for the captial stock and for man hours are taken from the database of the Deutsches Institut für Wirtschaftsforschung (DIW) mentioned above (cf. above definition 6. Firm size).

9. Wage level. The data for wages per man hour are computed from the database of the DIW mentioned above (cf. above definition 6. Firm size).

10. Tariffs. The data for the total protection in Germany (effective rate of protection plus subsidies) are taken from: Doris Witteler, "Tarifäre und nichttarifäre Handelshemmnisse in der Bundesrepublik Deutschland – Ausmaβ und Ursachen" in *Die Weltwirtschaft*, 1/1986, pp. 136–55, Table 2.

11. Private consumption. The data for the percentage of shipments delivered for private consumption are taken from: Statistisches Bundesamt, Fachserie 18, Reihe 2, *Input-Output Tabellen 1982*, Wiesbaden: Kohlhammer 1987.

The data for 1 to 9 are for 1983; the data for 10 to 11 are for 1982.

US data

The data of 247 four-digit SIC industries were taken for this study. The union participation variable and the R&D data is only based on a three-digit SIC level.

1. Innovation. The measure of innovative activity is from the US Small Business Administration and contains over 8,000 innovations identified in the sections of over 100 technology, engineering and trade journals. While the innovations were recorded in 1982, they were, in fact, the results of inventions made around 1978.

2. Concentration. The 1977 concentration ratio comes from the US Department of Commerce, Bureau of the Census, *Annual Survey of Manufacturers, Industrial Profiles* (issued by the US Government Printing Office in 1981).

3. Advertising. The data are from the US Input–Output Table.

4. Human capital. The labor skill measure is from the US Department of Commerce, Bureau of the Census, *Census Population 1970*, Occupation by Industry (issued by the US Government Printing Office in 1972).

5. Union influence. The data are from R.B. Freeman and J.L. Medoff (1979), "New estimates of private sector unionism in the United States", *Industrial and Labor Relations Review*, vol. 32, pp. 143–74.

6. Firm size. The large firm dummy is from the US Small Business Administration, Office of Advocacy, Small Business Database, *US Enterprise and Establishment Microdata* (USEEM) *file*.

7. R&D. The data were constructed by the United States International Trade Commission and are reported in US International Trade Commission, *Industrial Characteristics and Trade Performance Data Base*, Washington DC (isued 1975).

8. Capital intensity. See 2. Concentration.
9. Wage level. See 2. Concentration.
10. Tariffs. See 7. R&D.
11. Private consumption. We have created a dummy by defining the following SIC industries of those industries included in our sample as consumption industries:

> 2011–2259, 2291–2292, 2391–2395, 2397–2399, 2511–2519, 2647–2731, 2841–2842, 3142–3172, 3262–3263, 3421, 3432, 3631–3641, 3651–3652, 3692, 3711–3714, 3732, 3911–3952, 3961–3962, 3991, 3996–3999.

Notes

1. See Schmalensee (1986).
2. As far as we know only Lunn (1986), Hula (1988) and Schulenburg (1988) have focused recently on empirical study on the inter-relationship of advertising and innovation.
3. E.g. Kaldor (1950).
4. See for a theoretical treatment of this problem Schmalensee (1986).
5. See Kamien and Schwartz (1975), Scherer   1983), Pakes (1985), Baldwin and Scott (1987) and Acs and Audretsch (1988).
6. Stiglitz and Mathewson (1986).
7. A survey on empirical studies on advertising is presented by Farris and Albion (1981).
8. See Nelson (1974).
9. Buxton *et al.* (1984), however, claim that there exists a reverse quadratic U-shaped relationship between concentration and advertising intensity.
10. See e.g. Finsinger and Schulenburg (1987).
11. See Scherer (1965, 1980), Kamien and Schwartz (1975), Baldwin and Scott (1987) and Acs and Audretsch (1988). See for German studies König and Zimmermann (1986), Entorf (1988) and Pohlmeier (1988).
12. See Mansfield (1981) and Connolly and Hirschey (1984).
13. Hula (1988) explains the positive effect of advertising on innovation by expanded new product demands due to advertising.
14. A model for a counterexample is presented by Grossman and Shapiro (1984).
15. See also Hirsch and Link (1986), Addison and Hirsch (1989) and Hirsch and Connolly (1987). An alternative theory predicting a negative influence of unions on innovation was developed by FitzRoy and Kraft (1990).
16. See Baldwin and Scott (1987).
17. See Dasgupta (1986).
18. See Scherer (1980).
19. Krugman (1986, p.11), Stykolt and Eastman (1960) and Dixit and Norman (1980).
20. Lunn (1986) and Farber (1981) have tested similar simultaneous three-equation models. Lunn's endogenous variables are process or product innovations, concentration and advertising. Farber explores the inter-

relationship of advertising, concentration and employment of scientists and engineers.
21. cf. e.g. Rousseeuw and Leroy (1987).
22. cf. Cook and Weisberg (1982), p. 20 ff.
23. cf. White (1980).
24. See Spence (1986).
25. In a RESET version, cf. Nakamura and Nakamura (1981).
26. It should be pointed out, again, that differences in the definition of the US and FRG variables might influence the international comparisons.
27. cf. Wagner and Bellmann (1987).
28. cf. Wagner (1988).
29. cf. Wagner (1987).
30. cf. Scherer (1965).
31. cf. Chapter 2 of this book.
32. Methods for the analysis of panel data are discussed in detail by Hsiao (1986).
33. cf. Geroski (1987).

References

Acs, Z. J. and Audretsch, D. B. (1987), "Innovation, market structure and firm size", *Review of Economics and Statistics*, November, vol. 69, pp. 567–75.

Acs, Z. J. and Audretsch, D. B. (1988), "Innovation in large and small firms: An empirical analysis", *American Economic Review*, September, vol. 78, pp. 678–90.

Addison, J. T. and Hirsch, B. T. (1989), "Union effects on productivity, profits and growth: Has the long run arrived?", *Journal of Labor Economics*, vol. 7, pp. 72–105.

Arrow, K. (1962), "Economic welfare and the allocation of resources for invention" in R. R. Nelson (ed.), *The Rate and Direction of Inventive Activity*, Princeton NJ: Princeton University Press, pp. 609–25.

Audretsch, D.B. and Schulenburg, J.-M. Graf von der (1990), "Union participation, innovation and concentration: Results of a simultaneous model", *Journal of Institutional and Theoretical Economics*, vol. 146, pp. 298–313.

Baldwin, W. L. and Scott, J. T. (1987), *Market Structure and Technological Change*, London and New York: Harwood Academic Publishers.

Butters, G. R. (1977), "Equilibrium distribution of prices and advertising", *Review of Economic Studies*, October, vol. 44, pp. 465–92.

Buxton, A. J., Davies, S. W. and Lyons, B. R. (1984), "Concentration and advertising in consumer and producer markets", *Journal of Industrial Economics*, vol. 32, pp. 451–64.

Connolly, R. A. and Hirschey, M. (1984), "R&D, market structure and profits: A value based approach", *Review of Economics and Statistics*, November, vol. 66, pp. 682–6.

Connolly, R. A., Hirsch, B. T. and Hirschey, M. (1986), "Union rent seeking, intangible capital and market value of the firm", *Review of Economics and Statistics*, November, vol. 68, pp. 567–77.

Cook, R. D. and Weisberg, S. (1982), *Residuals and Influence in Regression*, London: Chapman & Hall.

Dasgupta, P. (1986), "The theory of technological competition" in J. Stiglitz and F. F. Mathewson (eds), *New Developments in the Analysis of Market Structure*, Cambridge, MA: MIT Press, pp. 519–47.

Dixit, A. K. and Norman, V. (1978), "Advertising and welfare", *Bell Journal of Economics*, Spring, vol. 9, pp. 1–18.

Dixit, A. K. and Norman, V. (1980), *Theory of International Trade*, Cambridge: Cambridge University Press.

Duetsch, L. L. (1984), "Entry and the extent of multiplant operations", *Journal of Industrial Economics*, June, vol. 32, pp. 477–87.

Entorf, H. (1988), "Die endogene Innovation. Eine mikro-empirische Analyse von Produktphasen als Innovationsindikatoren", *Jahrbücher für National-ökonomie und Statistik*, 1988, vol. 204, pp. 175–89.

Farber, J. (1981), "Buyer market structure and R&D effort: A simultaneous equations model", *Review of Economics and Statistics*, vol. 63, pp. 336–45.

Farris, P. and Albion, M. S. (1981), "Determinants of the advertising-to-sales ratio", *Journal of Advertising Research*, January, vol. 21, pp. 19–27.

Finsinger, J. and Schulenburg, J.-M. Graf von der (1987), "Nachfragever-halten bei unvollständigen Preisinformationen", *Jahrbücher für National-ökonomie und Statistik*, May, vol. 203, pp. 244–56.

FitzRoy, F. R. and Kraft, K. (1990), "Innovation, rent-sharing and the organization of labor," *Small Business Economics*, vol. 2, pp. 95–104.

Geroski, P. (1987), "Innovation, technological opportunity, and market structure", London Business School and University of Southampton (mimeo).

Grossman, G. M. and Shapiro, C. (1984), "Informative advertising with differentiated products", *Review of Economic Studies*, January, vol. 51, pp. 63–81.

Hausman, J. A. (1978), "Specification tests in econometrics", *Econometrica*, vol. 46, pp. 1251–71.

Hirsch, B. I. and Connolly, R. A. (1987), "Do unions capture monopoly profits", *Industrial and Labor Relations Review*, October, vol. 41, pp. 118–36.

Hirsch, B. and Link, A. N. (1986), "Labor union effects on innovative activity", *Journal of Labor Research*, vol. 8, pp. 323–32.

Hsiao, C. (1986), *Analysis of Panel Data*, Cambridge: Cambridge University Press.

Hula, D. G. (1988), "Advertising, new product profit expectations, and the firm's R&D investment decisions", *Applied Economics*, vol. 20, pp. 125–42.

Kaldor, N. (1950), "The economic aspects of advertising", *Review of Economic Studies*, January, vol. 18, pp. 1–27.

Kamien, M. I. and Schwartz, N. L. (1975), "Market structure and innovation: A survey", *Journal of Economic Literature*, March, vol. 13, pp. 1–37.

Khemani, R. S. and Shapiro, D. M. (1986), "The determinants of new plant entry in Canada", *Applied Economics*, November, vol. 18, pp. 1243–57.

König, H. and Zimmermann, K. F. (1986), "Innovations, market structure and market dynamics", *Journal of Institutional and Theoretical Economics*, vol. 142, pp. 184–99.

Krugman, P. (1986), "Industrial organisation and international trade", National Bureau of Economic Research (NBER) Working Paper, No. 1957, Cambridge, MA.

Lunn, J. (1986), "An empirical analysis of process and product patenting: A simultaneous equation framework", *Journal of Industrial Economics*, vol. 34, pp. 319–30.

Mansfield, E. (1981), "Composition of R and D expenditures: Relationship to size of firm, concentration and innovative output", *Review of Economics and Statistics*, November, vol. 63, pp. 601–15.

Nakamura, A. and Nakamura, M. (1981), "On the relationships among several specification tests presented by Durbin, Wu and Hausman", *Econometrica*, vol. 49, pp. 1583–8.

Orr, D. (1974), "The determinants of entry: A study of the Canadian manufacturing industries", *Review of Economics and Statistics*, February, vol. 56, pp. 58–676.

Pakes, A. (1985), "On patents, R&D and the stock market rate of return", *Journal of Political Economy*, April, vol. 93, pp. 390–409.

Pohlmeier, W. (1988), "On the determinants of innovative activity and market structure: Does simultaneity matter?", Institut für Volkswirtschaftlehre und Statistik der Universität Mannheim, Discussion Paper, 373–88, Mannheim: Universität Mannheim.

Rousseeuw, P. J. and Leroy, A. M. (1987), *Robust Regression and Outlier Detection*, New York: Wiley.

Scherer, F. M. (1965), "Firm size, market structure, opportunity, and the output of patented inventions", *American Economic Review*, December, vol. 55, pp. 1097–125.

Scherer, F. M. (1980), *Industrial Market Structure and Economic Performance*, Chicago, IL: Rand McNally College Publishing Co.

Scherer, F. M. (1983), "The propensity to patent", *International Journal of Industrial Organization*, March, vol. 1, pp. 107–28.

Scherer, F. M. (1990), "Changing perspectives on the firm size problem". See Chapter 2, this volume.

Schmalensee, R. (1986), "Advertising and market structure" in J. Stiglitz and G. F. Mathewson (eds), *New Developments in the Analysis of Market Structure*, Cambridge, MA: MIT Press, pp. 373–96.

Schulenburg, J.-M. Graf von der (1988), "Innovation, Marktstruktur und Werbung", *Hamburger Jahrbuch für Wirtschafts- und Gesellschaftspolitik*, vol. 33, pp. 141–53.

Schumpeter, J. A. (1950), *Capitalism, Socialism and Democracy*, 3rd edition, New York: Harper and Row.

Spence, M. (1986), "Cost reduction, competition and industry performance" in J. E. Stiglitz and G. F. Mathewson (eds), *New Developments in the Analysis of Market Structure*, Cambridge, MA: MIT Press, pp. 475–515

Stiglitz, J. E. (1986), "Theory of competition, incentives and risk" in J. E. Stiglitz and G. F. Mathewson (eds), *New Developments in the Analysis of Market Structure*, Cambridge, MA: MIT Press, pp. 399–446.

Stiglitz, J. E, and Mathewson, G. F. (eds) (1986), *New Developments in the Analysis of Market Structure*, Cambridge, MA: MIT Press.

Stykolt, S. and Eastman, H. (1960), "A model for the study of protected oligopolies", *Economic Journal*, June, vol. 70, pp. 336–47.

Telser, L. G. (1964), "Advertising and competition", *Journal of Political Economy*, December, vol. 72, pp. 537–62.

Wagner, J. (1987), "Zur politischen Ökonomie der Protektion in der Bundes-republik Deutschland", *Kyklos*, vol. 40, pp. 548–67.

Wagner, J. (1988), "Innovationen und Entwicklung der internationalen Wett bewerbsfähigkeit", *Jahrbücher für Nationalökonomie und Statistik*, vol. 205, pp. 50–64.

Wagner, J. and Bellmann, L. (1987), "Produkt- und Prozeβinnovationen als Unternehmensstrategien bei Importdruck", *IFO-Studien*, vol. 33, pp. 223–42.

White, H. (1980), "A heteroscedasticity-consistent convariance matrix estimator and a direct test for heteroscedasticity", *Econometrica*, vol. 48, pp. 817–38.

12

Firm size versus diversity in the achievement of technological advance*

Wesley M. Cohen and Steven Klepper

12.1 Introduction

Since the writings of Schumpeter, the role of firm size in promoting technical advance has preoccupied scholars of technological change. As international competitive pressures have intensified in recent years, the role of firm size has emerged as a central concern of policy-makers and industrialists as well. In the United States, some argue that the small entrepreneurial firm is the primary vehicle through which new ideas are introduced into the marketplace and that the diversity of ideas and approaches flowing from small firms represents American manufacturing's key competitive advantage, e.g. Gilder (1988), Rodgers (1990), Shaffer (1990). On the other side, it is argued that only large firms or consortia can command the resources necessary to field the large research efforts required to keep up with the large, often cooperatively organized research operations of Japan, and, more recently, Europe, e.g. Ferguson (1988), Norris (1983), Noyce (1990). Over the course of his career, Schumpeter himself spanned the poles of this debate. In his earlier work, Schumpeter (1934) saw the small scale entrepreneur as the key to capitalism's vitality. Later Schumpeter (1942) argued that the large scale enterprise was the principal engine of technological progress, although he feared that the bureaucratic character of the large modern corporation would undermine entrepreneurial initiative and eventually sap capitalism of its technological vitality.

In recent years, this controversy has been dominated, with some

* We are indebted to Mark Kamlet and Jonathan Leland for their suggestions.

exceptions, by industrialists and policy analysts. Most of the arguments advanced by both sides are based upon specific industries, most notably the semiconductor and computer industries. These arguments have largely failed to galvanize economists in particular. Not only have the arguments been casually developed, but there are strong counterarguments to each, leaving us with little *a priori* basis for favoring either position. For example, proponents of large scale enterprise often allude to scale economies and capital market imperfections to rationalize the need for large firms and consortia. Scale economies, however, bear only on the optimal size of the R&D effort and relate at best only indirectly to the optimal size of the firm. With regard to capital market imperfections, there is no consensus among economists as to the import of such imperfections for investment in general, much less R&D in particular. Those arguing on behalf of small size have claimed that small firms, unencumbered by bureaucracy, provide both the freedom and economic incentives that stimulate creativity and agility in response to economic opportunity. On the other hand, while large firms may indeed be more bureaucratic, they do, however, provide a superior human and capital infrastructure to support innovative activity.

To the extent that economists have probed the relationship between firm size and innovative activity, the evidence for either side of the debate has also not been compelling. Indeed, the predominant finding from a voluminous literature on the subject is that in most industries, above a modest threshold firm size, large firms are no more research intensive than smaller firms. This implies that consolidation of firms above the threshold would have no effect on total industry R&D expenditures, which has been widely interpreted to indicate that above a modest firm size neither small nor large firm size confers any advantage in R&D, e.g. Kamien and Schwartz (1982), Scherer (1980), Baldwin and Scott (1987).

We propose that even if small and large firms are equally capable at R&D, there may nonetheless be important social advantages associated with both large and small firm size. As such, we suggest that there is merit to both sides of the debate about the optimal firm size for R&D, but not for the reasons typically advanced. Our demonstration of the social advantages of small and large firm size proceeds in two stages. First, drawing heavily on prior work (Cohen and Klepper, 1989a), we suggest that both firm size and technological diversity influence firm R&D expenditures within industries and we highlight evidence that supports this claim. In the second stage, we go beyond our earlier analysis, and, in a more speculative spirit, consider not only how firm size and diversity influence firm R&D expenditures, but also how they affect the technical advance generated by the expenditures. We argue there are virtues both to having a large number of small firms in an industry and to consolidating output in a few large firms.

The rationale in our framework for the advantages of small scale enterprise rests on two notions suggested by Nelson's (1981) argument about the importance of competition and diversity for technological change. The first notion is that in the typical industry undergoing technological change, there are many productive ways of innovating. The other notion is that firms have different capabilities and perceptions about the approaches that are worth pursuing which lead them to pursue different sets of the available approaches to innovation. In such a world, dividing up industry output over a greater number of small firms increases the chances that any given approach to innovation will be pursued, thereby increasing the diversity of technological efforts in the industry. While increasing the number of firms does not necessarily benefit individual firms in the industry, it benefits society by increasing the number of productive approaches to innovation that are collectively pursued in the industry. From this perspective, the source of the social advantage associated with small firm size is not smallness *per se*, but the greater number of firms that small size implies given some industry demand.

Our argument about the social advantages of large firm size is based on the idea that pervasive imperfections in the market for information provide large firms with an advantage in appropriating the returns from innovation. As a result, the returns to large firms from any given R&D effort will be greater than the returns to small firms, leading large firms to conduct more socially desirable R&D. Consequently, consolidating industry output over a smaller number of firms will result in the performance of more socially desirable R&D for each approach to innovation that is pursued. This benefit is realized independent of whether there are any economies of scale in the conduct of R&D or whether large firms have superior ability to finance R&D.

Given the social advantages associated with both large firm size and more numerous small firms, there will always be a tradeoff associated with changing the number of firms within an industry. Reducing the number of firms will increase the average firm size, which will increase the intensity of innovative effort in each approach to innovation that is pursued. This comes, however, at the cost of reducing the number of productive approaches to innovation that are collectively pursued in the industry. On the other hand, increasing the number of firms will increase the diversity of approaches that are pursued in the industry but reduce the intensity of effort in each approach. The optimal number of firms in each industry will depend on the relative importance of these two effects, which we argue will vary across industries.

We speculate that market forces alone are not likely to lead to the socially optimal average firm size and number of firms in each industry. For example, the eventual displacement of larger incumbent firms by smaller new firms will compromise the appropriability advantages of size. Alternatively, even when entry of new firms can enhance social welfare by increasing the diversity of approaches to innovation pursued within an industry, we argue that it will not necessarily be forthcoming. Although these opposing forces could conceivably balance out to yield some optimal number of firms, we have no reason to expect this to occur in the typical industry.

We discuss how the failure of private initiatives to lead to a socially optimal state will create a tension from a policy standpoint. Following recent trends, obstacles to cooperative R&D efforts within industries could be removed, which would encourage the development of larger scale innovating units. This will, however, reduce the diversity of technological approaches pursued. Alternatively, diversity could be stimulated by government subsidies to entrants. However, more entry will reduce average firm size, thereby compromising the advantages of size. We suggest that it may well be just this tension that has given rise to the recent policy debates. We argue there are policies, however, that can to some degree preserve the benefits of size while at the same time maintaining technological diversity.

The chapter is organized as follows. In Section 12.2, we present our framework and briefly review the evidence that supports it. In Section 12.3, we suggest how larger firm size is tied to technical advance. In Section 12.4, we suggest how a larger number of firms may yield greater technological diversity, and how that diversity, in turn, contributes to technical advance. In Section 12.5, we discuss the tradeoff implied by the existence of both an appropriability advantage of large firm size and a diversity-inducing advantage of having numerous small firms within an industry. In Section 12.6, we conclude by highlighting the limitations of our analysis and by suggesting issues for further research.

12.2   A model of R&D investment

In this section we consider how firms decide how much to spend on R&D. Our conception follows the model of R&D investment developed in Cohen and Klepper (1989a). We discuss the basis for the model and provide a brief description of its implications. We refer the reader to Cohen and Klepper (1989a) for the formal and more extensive development of the model.

The model is based on a distinctive conception of an industry. The industry is assumed to be subject to ongoing technological change. Following Nelson (1990) and others, it is assumed that technological advance can be achieved in many different ways. A distinction is commonly drawn, for example, between process and product innovations. Moreover, there are typically many different ways of improving technology within each of these categories. For example, a production process may be improved through automation and by increasing the scale of production. Similarly, a product may be improved along a range of dimensions. For example, personal computers can be made more user-friendly and they can be made to work faster. These different approaches to enhancing product and process performance are often additive in their effect on technical advance; they do not represent competing or mutually exclusive approaches to satisfying the same performance objective. A firm in principle could pursue all of the approaches or any subset of them.

Firms are assumed to be uncertain about the profitability of pursuing any given approach to innovation, both in terms of the cost of the innovation and its gross returns. For simplicity, we assume that all approaches are profitable to pursue, although the firms do not know that in advance. The uncertainty concerning the profitability of the various approaches to innovation generates differences in the expectations of firms about which approaches are worth pursuing. These differences are assumed to reflect the idiosyncratic characteristics of individual decision-makers and the differences in the sorts of expertise accumulated by each firm due to its prior decisions. The differing expectations cause the firms to make different bets on which approaches to pursue.

The firms that make the best bets will experience the greatest rate of innovation and technical advance. It is assumed that eventually all innovations (but not the approaches themselves) will be copied, so that ultimately all benefits from innovation will be passed on to buyers in the form of lower prices. In the interim, however, firms that innovate most will reap economic profits. They will grow at the fastest rates over time, although growth will tend to be incremental, reflecting, among other things, the impact of uncertainty on investment and convex adjustment costs. Those firms that innovate the least will find the price of their product falling below their costs and will eventually exit. Over time the expertise required to evaluate and exploit approaches to innovation will change as the industry's technology evolves. Firms will be limited in their ability to change their expertise, and in the long run new firms with more suitable expertise will replace the industry leaders.

The consequence of this process is that at a given moment in each industry there will be an array of firms of different sizes and capabilities with different views about which approaches to innovation are worth pursuing. The model focuses on how these differences condition firm decisions about which approaches to pursue and about how much to spend on R&D for each approach pursued. The number of firms and the output of each firm are taken as given, having been determined by the evolutionary process leading up to that moment in the industry's history. Firms are also assumed to be price-takers and to choose their R&D expenditures independent of the R&D expenditures of their rivals.

With regard to the approaches a firm decides to pursue, we assume this choice is determined exclusively by the expertise of the firm. A firm is assumed to pursue a particular approach if it possesses the expertise that would enable it to exploit and recognize the value of the approach. Consequently, in each industry the environment governing the endowment of firms with expertise will ultimately determine the approaches to innovation that are pursued by firms in the industry. We use a very simple model to represent this environment. First, we assume that in each industry the likelihood of a firm being endowed with the expertise that would lead it to pursue any given approach is independent of the likelihood of the firm being endowed with the expertise that would lead it to pursue any other approach. This implies that the probability of a firm pursuing any given approach is independent of the probability of it pursuing any other. Second, we assume that in each industry the likelihood of a firm being endowed with the expertise that would lead it to pursue any given approach is independent of the size of the firm. This implies that the probability of a firm pursuing any given approach is independent of its size.

The choice about how much the firm should spend on each approach to innovation it pursues is conceived as follows. Each approach to innovation is assumed to be characterized by a range of projects with different marginal products, where the marginal product of a project is defined in terms of the technical advance (measured in standard units) generated from the project. Ordering these projects by their marginal products defines a diminishing marginal product schedule that relates the marginal product of R&D expenditures on an approach to the level of R&D expenditures on the approach, where it is assumed that the level of R&D expenditures is proportional to the number of projects pursued. All firms that pursue an approach are assumed to face the same marginal product of R&D schedule for the approach.

Although the marginal product schedules for a given approach are assumed identical across firms, the marginal revenue schedules are not. The marginal revenue a firm earns from a project equals the marginal product of the project times the level of output over which the innovations from the project are applied. It is assumed that imperfections in the market for information impede the sale of innovations in disembodied form (cf. Arrow, 1962). This implies that firms will appropriate the returns to their innovations primarily through their own output.[1] Since we also assume firm growth to be incremental, the returns a firm earns from any given R&D project will therefore be proportional to its sales prior to the R&D project.

The assumptions governing the choice of approaches to innovation and the amount spent per approach form the basis for predictions about firm R&D expenditures in an industry. First, assuming no economies of scale in R&D, Cohen and Klepper (1989a) demonstrate that the assumption that firms appropriate the returns to R&D through their own output implies that in each industry all firms pursuing a particular approach will spend an amount on R&D on the approach that is proportional to their sales. The simple intuition behind this result is that the greater the firm's sales then the larger the returns from any given R&D effort and thus the larger the optimal R&D expenditures of the firm. Furthermore, if the number of approaches to innovation pursued by the firm is uncorrelated with its sales, Cohen and Klepper (1989b) show that the overall R&D intensity of the firm (i.e. the ratio of the firm's R&D expenditures on all approaches to its sales) will be independent of its sales. Alternatively expressed, in each industry total firm R&D expenditures will vary across firms proportionally with the sales of the firms.

Second, Cohen and Klepper (1989a) demonstrate that if it is assumed that in each industry the marginal product schedule of each approach to innovation is the same and firms are equally likely to be endowed with the expertise that would lead them to adopt each approach to innovation, then the number of approaches to innovation pursued by firms and their overall R&D intensity will be binomially distributed. Intuitively, given that firms will pursue different numbers of approaches, the adoption of each approach by a firm can be thought of as an independent Bernouilli trial. As a consequence, in each industry the number of approaches pursued by firms will be binomially distributed. It was noted above that firms' R&D expenditures on any given approach will be proportional to their sales. Coupled with the assumption that the marginal product schedules of the approaches do not differ, Cohen and Klepper (1989a) show that this implies that in each industry firm R&D intensities will also be binomially distributed, with the firms that pursue the greatest number of approaches having the largest R&D intensities and the firms that pursue the smallest number of approaches having the smallest R&D intensities.

Both sets of predictions of the model are supported by evidence on how firm R&D expenditures vary within industries. Cohen and Klepper (1989b) note how the predictions of the model concerning the independence of firm R&D intensities and firm sales accords with the large body of evidence about the proportional relationship within industries between R&D expenditures and firm sales noted in the introduction. The predictions of the model concerning the nature of the distribution within industries of firm R&D intensities are supported by the empirical regularities in industry R&D intensity distributions documented by Cohen and Klepper (1989a) for a sample of business units[2] belonging largely to the 1,000 leading American manufacturing firms as of 1974–1977. These distributions tend to be unimodal, positively skewed, and to contain a substantial number of nonperformers of R&D, which is consistent with firm R&D intensity being binomially distributed. Moreover, across industries the moments of the distributions are correlated, which is also compatible with the binomial. Indeed, as Cohen and Klepper (1989a) show, the model is capable of explaining not only the fact that the moments are correlated, but also the signs of the correlations.

Thus, the model can explain a fairly exacting set of regularities that characterize the distribution of R&D intensities within industries. This provides indirect support for the two central tenets of the model:

(a)   returns to R&D are scaled by the size of the firm; and
(b)   differences in firm R&D intensities are attributable to differences across firms in the number of approaches to innovation pursued.

12.3   The advantages of large size

In this section we consider the implications of our framework for the advantages of large firm size. Our discussion is largely based on Cohen and Klepper (1989b). Following our earlier work, we assume that the mechanisms within an industry for appropriating rents due to innovation, such as patents, are given. We focus on the welfare implications of the link in our framework between firm size and the appropriability of the returns to innovation.

Since in our framework the firm earns returns on its innovations through its own output, the returns to R&D are scaled by the firm's sales. As we noted, this is the key in our framework to explaining the proportional relationship between R&D effort and firm size. This argument also implies that large firms will have an advantage over smaller firms in R&D competition. To see this, consider two firms that pursue the same number of approaches to innovation. Suppose, for simplicity, that all R&D expenditures result in innovations which lower the average cost of production.[3] If the two firms spend the same amount on R&D then our framework implies that they would achieve the same unit cost reduction. The larger firm, however, would apply the unit cost reduction over a larger level of output, enabling it to earn greater profits from its R&D. Moreover, our framework implies that if two firms pursue the same number of approaches, they will have the same R&D intensity, which implies that the bigger firm will actually perform more R&D than the smaller firm. This reflects the greater return it earns from any given level of R&D due to its greater size, which provides it with an incentive to perform a greater level of R&D than its smaller rival. Since it will earn positive profits from this additional R&D, this reinforces its advantage over its smaller rival.

Our interpretation of the relationship within industries between R&D effort and firm size is quite unconventional. The standard interpretation is that the proportional relationship between R&D effort and firm size suggests no advantage or disadvantage of large firms in R&D, although Fisher and Temin (1973) caution against making inferences about the advantages of firm size in R&D from the relationship between R&D effort and firm size. Of course, all we have shown is that the evidence is compatible with a theory in which there is an appropriability advantage to large firm size in R&D. However, Cohen and Klepper (1989b) note that there is no competing explanation for why R&D effort should be proportional to firm sales, and it is difficult to imagine an alternative explanation.

Our framework implies that the private advantage accruing to large firm size also represents a social advantage. Perhaps the easiest way to see this is to consider the merger of two firms that are the same size, pursue the same approaches to innovation, and spend the same amount on R&D on the same projects within each approach to innovation. The merger would result in two social dividends. First, duplicative R&D would be eliminated, a point suggested by other researchers (e.g. Spence, 1984). Second, and more novel, for each approach to innovation pursued by the two firms, the merger will increase the profitability of R&D *at the margin*, thereby leading the merged firm to undertake socially efficient R&D projects that were not undertaken by either firm before the merger. In effect, the merged firm will be better able to appropriate the returns from its innovations than either firm alone was able to do, causing it to undertake a greater number of socially efficient R&D projects than were collectively undertaken by the two firms before the merger.

Much of the empirical literature in the Schumpeterian tradition assumes that if R&D effort rises proportionally with firm size, then the size distribution of firms would have no effect on the amount of technological change achieved in the industry. In this view, technical advance is solely a function of the total amount of R&D performed in the industry, notwithstanding who performs it. In contrast, Cohen and Klepper (1989b) argue that if innovations cannot be sold in disembodied form, then the total amount of technological change generated in an industry will be a function of not only the total amount of R&D performed by all firms but also the distribution of R&D effort across firms. By redistributing sales and R&D expenditures from smaller to larger firms, total R&D expenditures would not change but a greater number of socially efficient R&D projects would be collectively undertaken and innovations from the projects would be applied to a greater portion of the industry's output, thereby increasing the amount of technological change generated per unit of industry R&D expenditure.

If large firms have an advantage in conducting R&D, as our framework implies, then the appropriability advantage of size would be fully realized under monopoly. We do not, however, commonly observe the emergence of monopoly in R&D intensive industries despite the appropriability advantage of size. An explanation for this is that even successful firms grow only slowly, and that over time technology typically evolves in ways that favor either new firms or firms that were formerly less successful. Indeed, in his development of the notion of creative destruction, Schumpeter (1942) suggested that even entrenched monopolists will be displaced in the long run for similar reasons.

In the absence of monopoly, firms will have an incentive to cooperate in R&D if they can appropriate the returns to their innovations only through their own output. For any given approach to innovation, all firms pursuing the approach could benefit by pooling their efforts. They would avoid the costly duplication of effort that would otherwise occur. Moreover, because they could apply innovations to their joint output, the firms would collectively profit from R&D expenditures that none of the firms could profit from individually. This suggests that it would be efficient for government to allow cooperative R&D ventures among competing firms pursuing the same approaches to innovation. Indeed, since 1984, US antitrust prohibitions against cooperative R&D ventures have been relaxed and numerous cooperative ventures have been initiated. A large fraction of these cooperative ventures, however, have brought together firms with complementary capabilities rather than firms pursuing the same approaches to innovation. This suggests that firms have a range of incentives to cooperate in their R&D efforts, and the one we have identified may not, in practice, be paramount. Moreover, cooperation among firms pursuing the same approaches to innovation may be impeded by difficulties in communication and concerns about competitive advantage. If so, then a more active role for government may be indicated.

It is important to stress that the advantage of size in our framework is based solely on the assumption that firms can appropriate the returns to their innovations only through their own output (prior to innovating). To the extent that innovations can be sold in disembodied form and/or that firms can reap the returns to their innovation through rapid growth, the advantage of (*ex ante*) large firm size will be less significant. It is likely that both of these factors will vary across industries and perhaps types of innovations (e.g. product versus process innovations).[4] The greater the extent to which the returns to innovation do not depend on *ex ante* output, then the lower will be the social benefits from cooperation.

12.4 The advantages of diversity

In this section we consider the role of technological diversity in technical advance and the link between diversity and the number of competitors.

Diversity is incorporated in our framework through the assumption that firms pursue different sets of approaches to innovation. This in turn gives rise to differences in firm R&D intensities that reflect differences in the total number of approaches to innovation pursued by each firm. As we discussed, this form of technological diversity allows us to explain regularities characterizing the nature of the distribution within industries of firm R&D intensities. Of course, this does not mean that diversity is actually an important determinant of R&D intensities within industries, but it does provide indirect support for this view.

In our framework, diversity stems from the existence of different approaches to innovation within an industry. We conceive of these different approaches not as competing or parallel ways of achieving the same technological objective but as independent ways of achieving technological change.[5] The total number of approaches to innovation in an industry can be thought of as being determined by the science and technology underlying R&D in the industry. Which of these approaches are actually pursued by firms in the industry at any given moment will depend upon the perceptions and capabilities of firms in the industry. It is certainly possible that some subset of the approaches will not be pursued by any firm in the industry. Indeed, if there are X firms in the industry and the probability of any approach being pursued by any given firm is p, then the likelihood of any approach not being pursued by any firm in the industry is $[1 - (1 - p)^X]$. A simple measure of technological diversity in our framework is the number of distinct approaches to innovation that are actually pursued by the firms in the industry. This measure corresponds well to the notion of diversity implicit in the arguments of proponents of small scale enterprise who stress the greater breadth of ideas and types of innovations pursued in industries composed of numerous small firms, e.g. Rodgers (1990).

It follows directly from this measure of technological diversity that the extent of diversity will be directly related to the number of firms in the industry. Since the likelihood of a potential approach to innovation in an industry not being pursued by any firm is $[1 - (1 - p)^X]$, the expected number of different approaches to innovation pursued in an industry will be a function of the number of firms, X. This implies that on average industries composed of a greater number of firms will be characterized by a greater amount of technological diversity. Note that this greater diversity does not emerge from any superior creativity on the part of smaller firms. It is simply the result of having a greater number of firms in the industry choosing which approaches to innovation to pursue.

The predicted relationship between the number of firms and technological diversity depends upon a key assumption of our framework, namely that the likelihood of a firm pursuing any given approach is independent of the size of the firm. If, alternatively, the probability of a firm pursuing any given approach increased with the size of the firm, it would no longer follow that an increase in the number of firms would always increase the expected degree of diversity.[6] While there are undoubtedly industries where the probability of pursuing an approach is correlated with firm size, the evidence concerning the regularities in the industry R&D intensity distributions suggests that the number of such industries is limited. Otherwise, our framework implies that there would be a large number of industries for which the largest firms are characterized by above average R&D intensities,[7] which does not conform with the evidence. Moreover, if the probability of pursuing an approach were an increasing function of the size of the firm, then based on our framework the distribution of firm R&D intensities within an industry would depend on the size distribution of firms and would, in all likelihood, not conform to the regularities in the industry R&D intensity distributions documented by Cohen and Klepper (1989a).

If the likelihood of pursuing an approach to innovation is independent of the size of the firm then two firms of equal size would together be expected to pursue a greater number of distinct approaches to innovation than one firm twice their size. Having more firms provides more "independent minds" to consider the alternatives, which results in greater technological diversity. There are a number of organizational factors that might explain why diversity is promoted by a greater number of firms. We suspect it has something to do with the way R&D proposals are processed within firms. Suppose that proposals are initiated by technical staff and then subsequently considered by higher level decision-makers. If large firms have more hierarchical levels than small firms and a proposed approach can be rejected at any level, an approach will have a greater chance of being approved by a smaller firm. As long as the number of proposed approaches in the two firms and the one large firm are equal, the two firms would then be expected to pursue a greater number of distinct approaches than the large firm. Alternatively, suppose the probability of an approach being rejected is independent of the size of the firm. Even in this case, if all approaches proposed in the large firm are proposed in at least one of the two smaller firms and some are proposed in both of the smaller firms,[8] then the two smaller firms combined will approve a greater number of distinct approaches than the larger firm.

Suppose we are correct and the number of firms is an important determinant of the diversity of approaches to innovation pursued in an industry. In order for there to be an advantage associated with small firm size, there must be some benefit associated with the greater diversity brought about by a greater number of firms. Our framework highlights an obvious benefit from the greater diversity – the exploitation of beneficial approaches to innovation that otherwise would not have been pursued. This does not mean, however, that social welfare is necessarily promoted by the greater diversity associated with an increase in the number of firms. Holding the output of the industry constant, if the number of firms is increased then the average size of each firm will decline. In the prior section, we showed how in our framework the consolidation of firms would yield social benefits. It follows that reducing the average firm size would impose a social cost. Assuming that firms cannot sell innovations and can benefit from them only by incorporating them in their own output, each firm will spend less on R&D on the approaches it pursues and the innovations developed by each firm will be applied over a smaller level of output. The net effect on social welfare of increasing the number of firms depends on the magnitude of these costs relative to the benefits of having additional approaches to innovation pursued. We consider this tradeoff further in the next section.

While it is not straightforward to establish the social advantage of diversity in our framework, it is possible to imagine circumstances in which the increase in diversity brought about by a greater number of firms unequivocally promotes social welfare. For instance, suppose that firms are fully able to sell their innovations and there is no duplication among firms in their R&D efforts. In such a world, all innovations will be applied to the entire output of the industry, and a reduction in the size of each firm will not diminish its incentives to conduct R&D. Then the decrease in firm size associated with the increase in the number of firms will have no effect on the R&D expenditures of the incumbent firms nor on the output over which any innovations resulting from these expenditures are applied. Consequently, any increase in the number of approaches pursued resulting from an increase in the number of firms will provide a net benefit to society. Intuitively, having a greater number of different minds (i.e. firms) evaluate the possible approaches to innovation will result in a lower chance that a beneficial approach to innovation will be overlooked.

Implicit in the discussion of diversity is that at any given moment there may be firms outside of an industry that would pursue an otherwise unpursued approach if they entered the industry. Stated more baldly, it assumes that there may be potential entrants who recognize an opportunity but who do not enter the industry to exploit it. One reason this could occur is if there is some cost to entering the industry that exceeds the gross profit the entrant could make from pursuing the unexploited opportunity. From a social welfare perspective, however, this cost should be counted against the technical advance generated by the increase in the number of firms. If the social benefit of this technical advance was solely limited to the producer surplus resulting from the entrant's pursuit of a new approach, the net change in social welfare from the increase in the number of firms would be negative since the cost of entry would exceed the gross profits the producer could make from entering. Thus, for the increase in diversity to bring about an increase in social welfare, there must be some kind of wedge between the private and social returns from the pursuit of a new approach.

Our framework provides the basis for such a wedge. In laying out the framework, we noted that eventually all innovations are imitated. To the extent that the costs of imitation are less than the benefits, the social returns to innovation will exceed the private returns.[9] Consequently, it is possible that the pursuit of new approaches to innovation by entrants could be socially beneficial but not profitable to the entrant. In that case, the social benefits from greater diversity would not be realized through private initiative.

Our arguments suggest that if policies to increase the number of firms in an industry were undertaken, diversity would be expected to increase, which, in turn, could increase social welfare. Such policies could include prohibitions on horizontal mergers, subsidies of failing firms, and/or subsidies of entrants. Note that none of these policies requires the government to have any special insights about the sort of innovations an industry should pursue. Indeed, there is no reason to expect the government to be any more enlightened than firms about the best innovations to pursue. The rationale for these policies is simply that by getting more firms in the industry, there will be less chance that beneficial opportunities for technical advance will be unexploited.

It is important to note that the diversity-inducing effect of the presence of many small firms clearly depends upon the vitality of the science and technology underlying technical advance in any given industry, which we take as given. If this science and technology base is moribund, then the number of approaches to innovation that could be pursued will be few. In this case, the relationship between the number of firms and technological diversity, and, in turn, technical advance, will be weaker.

12.5   The tradeoff

In the last section we argued that it is not possible to bring about an increase in diversity through an increase in the number of firms without compromising some of the benefits associated with large firm size. Not surprisingly, it is also not possible to reap the benefits of large firm size without compromising the advantages associated with greater diversity. In this section we consider further the nature of this tradeoff.

When diversity is promoted through a greater number of firms, average firm size is reduced, which compromises the advantages associated with large size firms. Consider, alternatively, the costs associated with promoting the advantages of large firm size through the consolidation of firms. By reducing the number of firms in the industry, the expected number of approaches to innovation collectively pursued in the industry will decline. Then, on average, a smaller fraction of the beneficial opportunities for innovation will be pursued, which will reduce social welfare. Thus, just as increasing diversity through an increase in the number of firms has offsetting costs, so does increasing firm size through the consolidation of firms.[10]

Thus, in our framework, there is a tradeoff between the advantages associated with small firms and those associated with large firms. In order to have more approaches to innovation pursued, it is necessary to sacrifice some intensity of effort for each approach, and vice versa. The nature of this tradeoff is complicated and will depend on a number of factors. Analyzing the optimal average firm size for any industry in our framework requires a formal model of this tradeoff, which is beyond the scope of this chapter. We can, however, use our framework to speculate about the role of two factors that would be expected to shape the nature of the tradeoff: the extent to which firms can sell their innovations and the vitality of an industry's technology, as represented by the number of different approaches to innovation potentially available in the industry. To the extent that firms can realize rents due to innovation via licensing and other mechanisms that do not exploit the firm's output, the appropriability advantage of size will be limited. In this case, an industry composed of a greater number of firms will be desirable. To the extent that the science and technology base of an industry is such that the potential number of approaches to innovation available to the industry is small, the diversity advantage associated with numerous small firms will be limited. In this case, an industry composed of a smaller number of firms will be desirable.

Consider the case of an industry in which there are a large number of approaches to innovation and firms can appropriate the returns to innovation only through their own output, which is the case where the tradeoff between the advantages associated with large and small firm size is the most acute. We suspect a substantial number of technologically progressive industries will be characterized by just such an acute tradeoff. There is no reason to believe that in these types of industries market forces alone will result in an optimal number of firms. In any given industry, there likely will be either too little entry to promote sufficient diversity or entry may be so unencumbered that the appropriability advantage of larger firm size is excessively compromised.

If market forces cannot be expected to lead to the optimal number of firms, policy-makers will be faced with the challenge of deciding whether the number of firms in each industry is too large or too small. This will require making extremely difficult judgments. It is possible, however, that we have overblown the challenge posed by the tradeoff and policies can be designed to achieve the benefits of both diversity and large size. Consider, for example, current proposals that government condone firm cooperation on "generic" research (e.g. Nelson, 1990). If "generic" research includes research activity that more than one firm pursues, then such proposals can be interpreted as suggesting that firms pursue separately R&D on approaches which they do not pursue in common and pursue together R&D on approaches which they do pursue in common. In this fashion, cooperative agreements could be formulated that would permit the industry and society to benefit from the appropriability advantage of large firm size without compromising the benefits of diversity. Indeed, cooperation on something resembling what we call generic research is apparently one of the models for cooperative research employed in the Japanese manufacturing sector.

12.6   Conclusion

Our analysis lends support to both sides of the debate concerning the optimal firm size for achieving technical advance. It provides a basis for why industries composed of many small firms will tend to exhibit greater diversity in the approaches to innovation pursued, and why greater diversity will contribute to more rapid technological change. It also provides a basis for why industries populated by larger firms will achieve a more rapid rate of technical advance on the approaches to innovation that are pursued. These arguments together suggest that a tradeoff exists between the appropriability advantage of large size and the advantages of diversity that accrue from numerous small firms. Others, such as Nelson (1981), have also recognized a tradeoff between the diversity-inducing advantage of more competitive industry structures and advantages of large firm size, but not the particular tradeoff we have identified.

Our analysis has been more appreciative than rigorous and, indeed, often explicitly speculative. While we attempted to raise important questions, our framework requires more structuring before we can be confident about any of our conclusions. Even in its inchoate form, however, our analysis demonstrates that much needs to be done before the current debate about firm size can seriously inform policy. If we accept the plausibility of our basic framework, it focuses attention on a range of issues and questions. The fundamental premise of our analysis is that firm capabilities and perceptions differ within industries. This premise is not, however, widely reflected in analyses of industry behavior and performance, which typically take some representative firm as their starting point. Indeed, the analytic utility of our particular premise deserves scrutiny. Are differences in firm capabilities and perceptions as critical to explaining the industry patterns in innovative activity and performance as we suggest? Do these differences persist? More to the point, is our abstract characterization of these differences and their effects on innovative activity up to the task of providing a strong basis for policy?

These intra-industry differences in capabilities and perceptions underpin the hypothesized relationship in our framework between the number of firms within an industry and the number of distinct technological activities pursued by the industry as a whole. Surely this hypothesis should be tested. To establish the relationship between numbers of firms and technological diversity, we also made two important assumptions, which themselves should be examined. First, we assume that firms independently decide upon which approaches to innovation to pursue. This assumption precludes the clustering of firms around innovative activities due to imitation, a phenomenon highlighted by Nelson (1981) and Scott (1990). To the degree that innovative activities yield relatively fast, public results, the assumption may be suspect. While our evidence indirectly suggests that such clustering may not be critical for explaining innovative activity in a wide range of industries, more research would be helpful. Second, we assume that the number of approaches to innovation pursued by firms is independent of their size, implying large and small firms will tend to pursue the same number of approaches. This assumption probably does not apply to the smallest firms within an industry, particularly to the extent that such firms are often not full line manufacturing firms. Does it apply, however, to the medium to large firms that account for the preponderance of R&D and economic activity in the manufacturing sector? While our evidence again provides indirect support for this claim, more empirical and theoretical research is indicated.

We also make numerous other claims and assumptions that deserve further attention. For example, we argue that greater technological diversity stimulates technical advance and provides gross increments to social welfare. Assuming it exists, the mechanism linking diversity and technical advance has never been examined empirically and is not obvious. Our assumption that firm growth is incremental plays an important role in permitting us to hypothesize an appropriability advantage of large size. Again, both the assumption and its alleged effect on innovative performance are worth examining. Finally, we also need to test whether the relationship between R&D and firm size within industries depends upon appropriability conditions, particularly upon the extent to which firms can sell their innovations.[11] In conclusion, this litany of reasonable but unsubstantiated assumptions and arguments should make clear that this chapter is only a modest beginning of a daunting research agenda.

Notes

1. The survey results of Levin *et al.* (1987) also provide a strong empirical basis for the claim. They find that the most effective mechanism for appropriating rents due to innovation are first-mover advantages, secrecy, and complementary sales and service, all of which are implemented via the firm's own sales.
2. A business unit represents a firm's activity in a given industry.
3. The following argument applies equally to product innovations (Cohen and Klepper, 1989a).
4. For example, the ability of firms to sell their innovations in disembodied

form will vary across industries according to the ease of defining and enforcing property rights over innovations.

5. Our stylized conception of technological diversity differs from how others have conceptualized diversity. For example, in the analyses of Evenson and Kislev (1976) and Nelson (1982), different approaches represent different ways of achieving the same technological end. Differentiating approaches in this way is relevant when there is uncertainty about how best to achieve a particular technological objective. In our framework, we have abstracted from this kind of uncertainty. Our firms differ on which objectives they think are worthwhile to pursue, not on how to pursue the objectives. While no doubt there is always uncertainty about how best to achieve a particular technological objective, we have abstracted from this kind of uncertainty because it would not readily explain the regularities in the industry R&D intensity distributions. Indeed, it does not readily explain why R&D intensities of firms in an industry should differ at all in any systematic way (Cohen and Klepper, 1989a).

6. Holding industry output constant, an increase in the number of firms will reduce the average firm size. If the probability of a firm pursuing an approach were positively related to its size, then the increase in the number of firms would decrease the probability of any firm pursuing any given approach. Then the probability of any given approach not being pursued by any firm in the industry could fall with an increase in the number of firms, which would cause the expected amount of diversity to fall.

7. In our framework, if the probability of pursuing an approach were positively related to the size of the firm, then larger firms would be expected to pursue a greater number of approaches to innovation, causing them to have above average R&D intensities.

8. This is plausible if some of the approaches proposed in the large firm are proposed by multiple individuals. In that case, if the larger firm were broken into two firms, there would be some probability that the technical personnel in each of the small firms would propose the same approach in common.

9. In a static setting, imitation reinforces the gross benefits accruing from greater diversity because the benefits from the pursuit of new approaches will be reaped by more than just the firms pursuing the new approaches. Imitation may also yield a dynamic dividend in that pursuing a given approach may stimulate other firms "to improve it, variegate it, more generally contribute to its further advance," (Nelson, 1989, p. 6) even in ways that cannot be anticipated when the approach is first pursued.

10. We managed to sidestep this cost in the example we used earlier to illustrate the benefits of large firm size by focusing on the merger of two firms that pursued the same approaches to innovation. This insured that the merger of the firms did not reduce the diversity of approaches pursued in the industry. However, firms generally will not pursue the identical set of approaches to innovation, so that consolidations will be expected to compromise the collective number of approaches to innovation pursued in the industry.

11. Cohen and Klepper (1989b) demonstrate that if firms can sell some fraction of their innovations in disembodied form then large firm size will confer less of an advantage and R&D effort should rise less than proportionally with firm size.

Reference

Arrow, K. (1962) "Economic welfare and the allocation of resources for inventions" in R. R. Nelson, ed., *The Rate and Direction of Inventive Activity*, Princeton, NJ: Princeton University Press for the National Bureau of Economic Research.

Baldwin, W. L. and J. T. Scott (1987) *Market Structure and Technological Change*, Chur: Harwood Academic Publishers.

Cohen, W. M. and S. Klepper (1989a) "The random character of innovative effort", mimeo, Pittsburgh: Carnegie Mellon University.

Cohen, W. M. and S. Klepper (1989b) "A reprise of size and R&D", mimeo, Pittsburgh: Carnegie Mellon University.

Evenson, R. E. and Y. Kislev (1976) "A stochastic model of applied research", *Journal of Political Economy*, vol. 84, pp. 265–81.

Ferguson, C. H. (1988) "From the people who brought you voodoo economics", *Harvard Business Review*, May–June, vol. 89, pp. 55–62.

Fisher, F. M. and P. Temin (1973) "Returns to scale in research and development: What does the Schumpeterian hypothesis imply?", *Journal of Political Economy*, vol. 81, pp. 56–70.

Gilder, G. (1988) "The revitalization of everything: The law of the microcosm", *Harvard Business Review*, March–April, vol. 88, pp. 49–61.

Kamien, M. I. and N. L. Schwartz (1982) *Market Structure and Innovation*, Cambridge: Cambridge University Press.

Levin, R. C., A. Klevorick, R. Nelson and S. Winter (1987) "Appropriating the returns from industrial research and development", *Brookings Papers on Economic Activity*, No. 3, pp. 783–820.

Nelson, R. R. (1981) "Assessing private enterprise: An exegesis of tangled doctrine", *Bell Journal of Economics*, vol. 12, pp. 93–111.

Nelson, R. R. (1982) "The role of knowledge in R&D efficiency", *Quarterly Journal of Economics*, vol. 97, pp. 453–470.

Nelson, R. R. (1989) "What is 'commercial' and what is 'public' about technology, and what should be?", mimeo, presented at the Conference on Economic Growth and the Commercialization of New Technologies, Stanford University, September 11–12, 1989.

Nelson, R. R. (1990) "Capitalism as an engine of progress", *Research Policy*, vol. 19, pp. 193–214.

Norris, W. C. (1983) "How to expand R&D cooperation", *Business Week*, April 11, p. 21.

Noyce, R. N. (1990) "Cooperation is the best way to beat Japan", *New York Times*, July 9, Section 3, p. 2.

Rodgers, T. J. (1990) "Landmark messages from the microcosm", *Harvard Business Review*, January–February, vol. 90, pp. 24–30.

Scherer, F. M. (1980) *Industrial Market Structure and Economic Performance*, 2nd edn, Chicago: Rand McNally College Publishing Company.

Schumpeter, J. A. (1934) *The Theory of Economic Development*, Cambridge, Mass.: Harvard University Press.

Schumpeter, J. A. (1942) *Capitalism, Socialism, and Democracy*, New York: Harper.

Scott, J. (1990) "Research diversity and technical change", in Z. Acs and D. Audretsch, eds, *Innovation and Technical Change*, Ann Arbor: University of Michigan Press.

Shaffer, R. A. (1990) "Let a thousand companies fight", *New York Times*, 9 July, Section 3, p. 2.

Spence, A. M. (1984) "Cost reduction, competition and industry performance", *Econometrica*, vol. 52, pp. 101–21.

Index